Student Edition

Eureka Math
Grade K
Module 4

Special thanks go to the Gordon A. Cain Center and to the Department of Mathematics at Louisiana State University for their support in the development of *Eureka Math*.

For a free *Eureka Math* Teacher
Resource Pack, Parent Tip
Sheets, and more please
visit www.Eureka.tools

Printed in the U.S.A.

This book may be purchased from the publisher at eureka-math.org

10 9 8

ISBN 978-1-63255-285-3

Name _____   Date _____

Draw the light butterflies in the number bond. Then, draw the dark butterflies. Show what happens when you put the butterflies together.

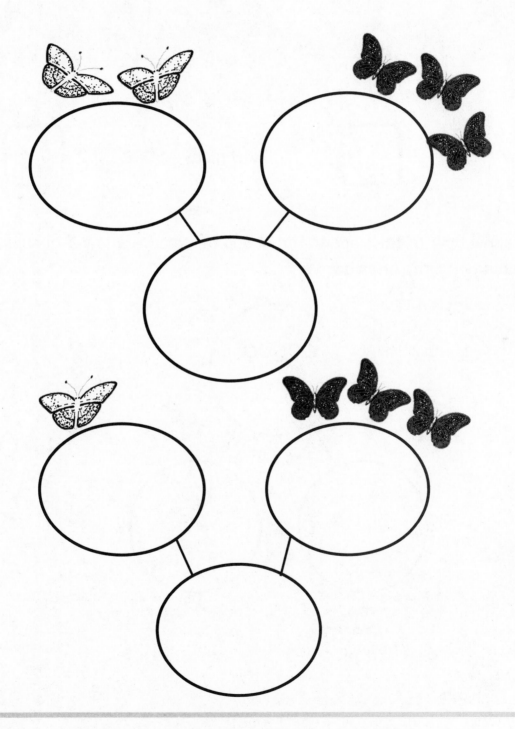

EUREKA MATH™   **Lesson 1:**   Model composition and decomposition of numbers to 5 using actions, objects, and drawings.

©2015 Great Minds. eureka-math.org
GK-M4-SE-B3-1.3.1-01.2016

1

Name _____   Date _____

How many 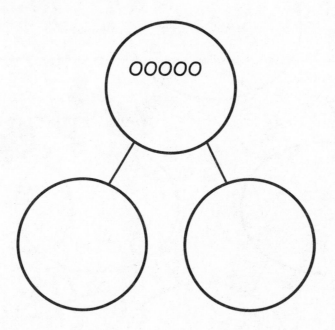 ? ⬜   How many 🐱 ? ⬜

Draw to show how to take apart the group of cats to show 2 groups, the ones sleeping and the ones awake.

Lesson 1:   Model composition and decomposition of numbers to 5 using actions, objects, and drawings.

©2015 Great Minds. eureka-math.org
GK-M4-SE-B3-1.3.1-01.2016

EUREKA MATH™

Name _____   Date _____

Draw the blue fish in the first circle on top. Draw the orange fish in the next circle on top. Draw all the fish in the bottom circle.

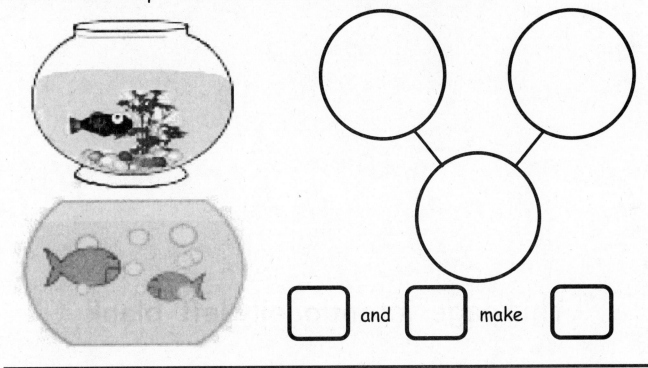

and   make

Draw a square for each fish in the top circle. Draw a square for each goldfish in the bottom circle. In the last circle on the bottom, draw a square for each spiny fish.

and   make

EUREKA MATH™

Lesson 1:   Model composition and decomposition of numbers to 5 using
             actions, objects, and drawings.

©2015 Great Minds. eureka-math.org
GK-M4-SE-B3-1.3.1-01.2016

3

This page intentionally left blank

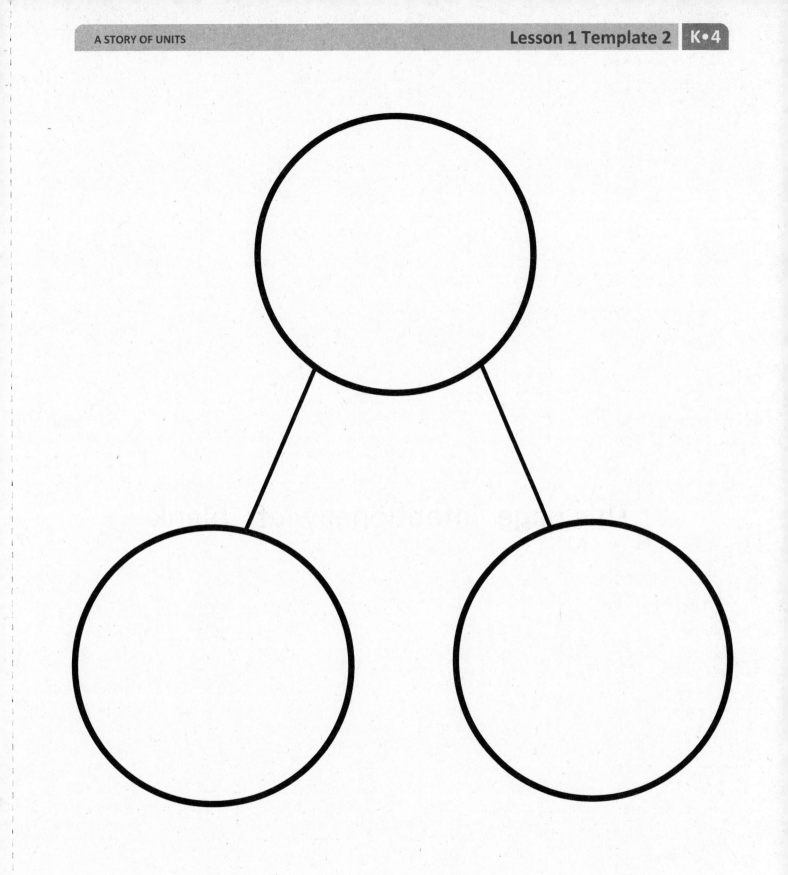

number bond

Lesson 1:   Model composition and decomposition of numbers to 5 using
actions, objects, and drawings.

©2015 Great Minds. eureka-math.org
GK-M4-SE-B3-1.3.1-01.2016

5

This page intentionally left blank

Name _____   Date _____

The squares below represent a cube stick. Color the squares to match the rabbits. 4 squares gray. 1 square black. Draw the squares in the number bond.

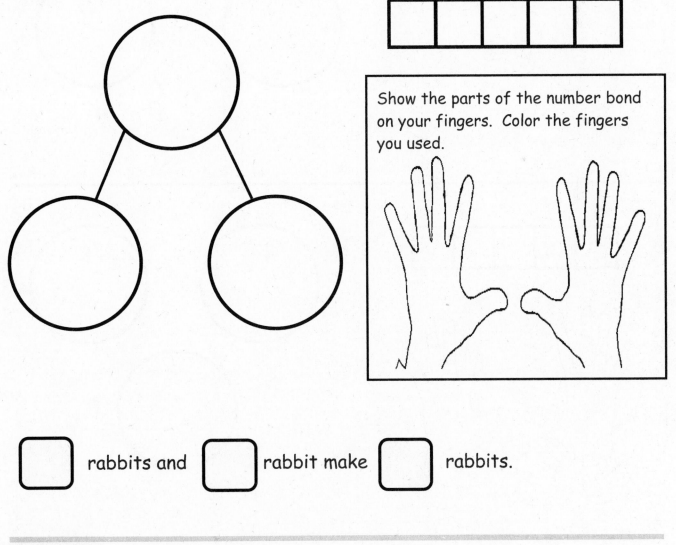

Show the parts of the number bond on your fingers. Color the fingers you used.

⬜ rabbits and ⬜ rabbit make ⬜ rabbits.

EUREKA MATH™

Lesson 2:   Model composition and decomposition of numbers to 5 using fingers and linking cube sticks.

©2015 Great Minds. eureka-math.org
GK-M4-SE-B3-1.3.1-01.2016

7

Name _____   Date _____

The squares below represent a cube stick. Color some squares blue and the rest of the squares red. Draw the squares you colored in the number bond. Show the hidden partners on your fingers to an adult. Color the fingers you showed.

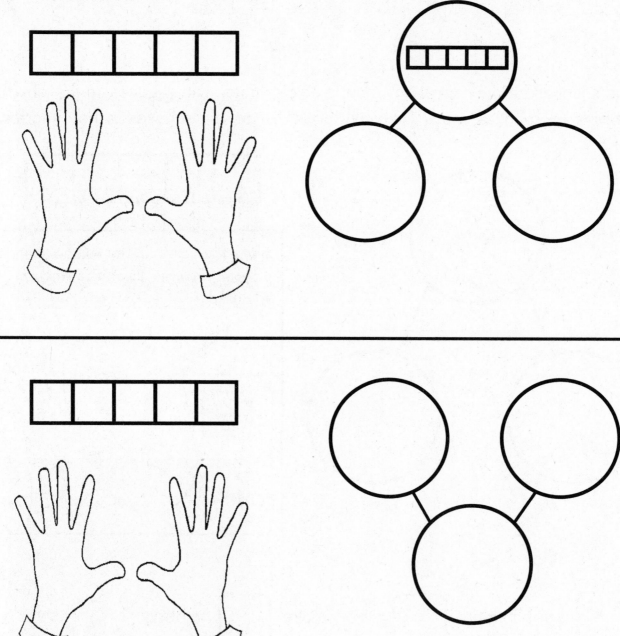

  Lesson 2:   Model composition and decomposition of numbers to 5 using fingers and linking cube sticks.

©2015 Great Minds. eureka-math.org
GK-M4-SE-B3-1.3.1-01.2016

EUREKA MATH™

Name _____   Date _____

Draw the shapes and write the numbers to complete the number bonds.

5

Lesson 3:   Represent composition story situations with drawings using numeric number bonds.

©2015 Great Minds. eureka-math.org
GK-M4-SE-B3-1.3.1-01.2016

Write numbers to complete the number bond. Put the number of dogs in one part and the number of balls in the other part.

Look at the picture. Tell a story about the birds going home to your neighbor. Draw a number bond, and write numbers that match your story.

Lesson 3: Represent composition story situations with drawings using numeric number bonds.

©2015 Great Minds. eureka-math.org
GK-M4-SE-B3-1.3.1-01.2016

Name _____   Date _____

Fill in the number bond to match the domino.

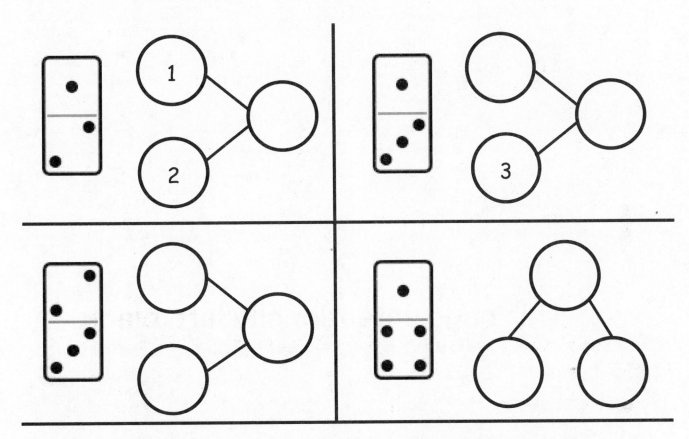

Fill in the domino with dots, and fill in the number bond to match.

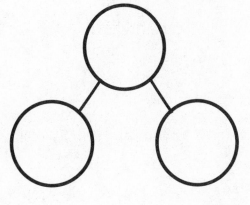

EUREKA
MATH™

Lesson 3:   Represent composition story situations with drawings using numeric
number bonds.

11

This page intentionally left blank

Name _____   Date _____

Draw and write the numbers to complete the number bonds.

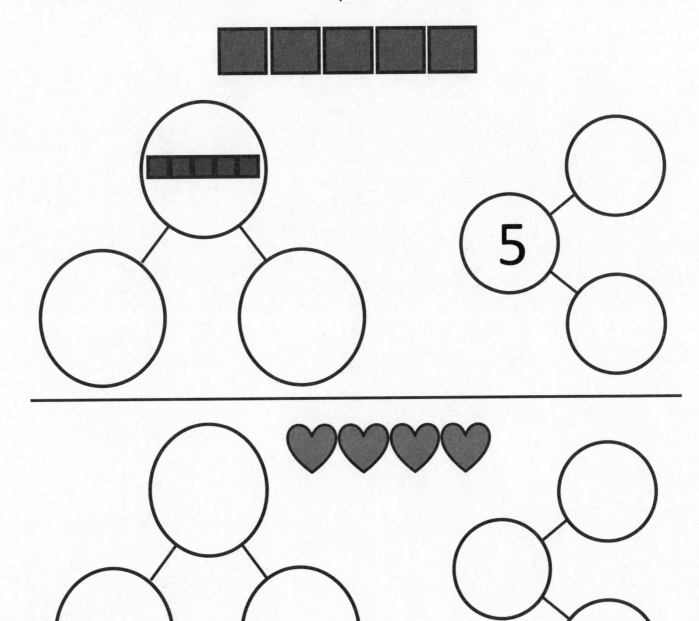

EUREKA
MATH™

Lesson 4:   Represent decomposition story situations with drawings using
numeric number bonds.

13

©2015 Great Minds. eureka-math.org
GK-M4-SE-B3-1.3.1-01.2016

Look at the picture. Tell your neighbor a story about the dogs standing and sitting. Draw a number bond, and write numbers that match your story.

Lesson 4:   Represent decomposition story situations with drawings using numeric number bonds.

Name _____   Date _____

Finish the number bonds. Finish the sentence.

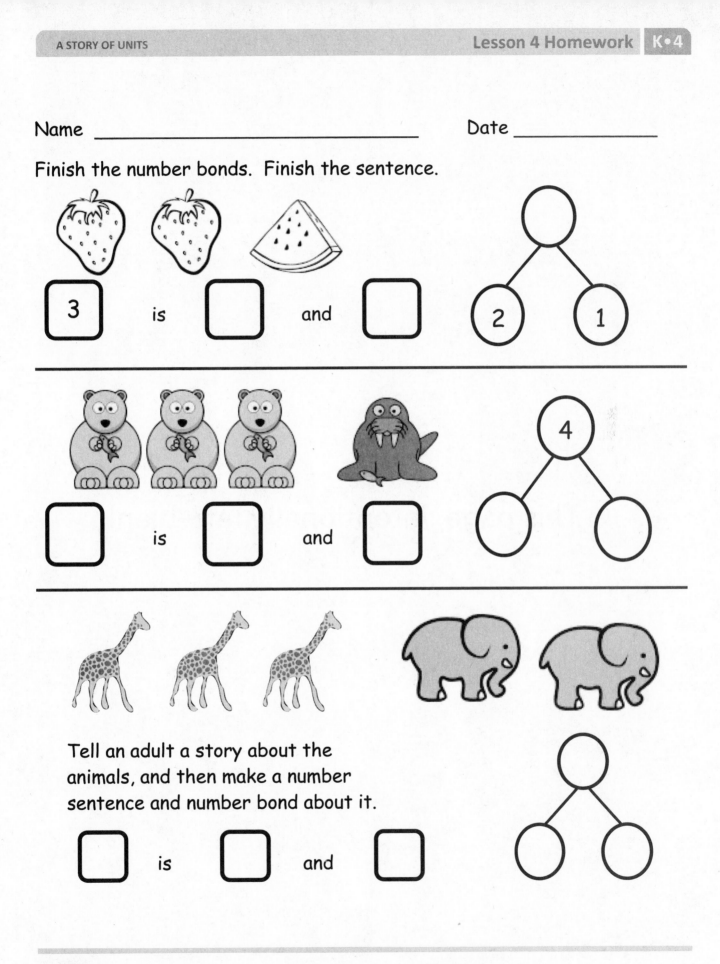

3   is   ☐   and   ☐

☐   is   ☐   and   ☐

Tell an adult a story about the animals, and then make a number sentence and number bond about it.

☐   is   ☐   and   ☐

Lesson 4: Represent decomposition story situations with drawings using numeric number bonds.

15

This page intentionally left blank

Name _____   Date _____

Write numbers to fill in the number bonds.

Lesson 5:   Represent composition and decomposition of numbers to 5 using pictorial and numeric number bonds.

©2015 Great Minds. eureka-math.org
GK-M4-SE-B3-1.3.1-01.2016

Name _____   Date _____

There are 2 pandas in a tree. 2 more are walking on the ground. How many pandas are there? Fill in the number bond and the sentence.

____ and ____ make ____

Tell a story about the penguins. Fill in the number bond and the sentence to match your story.

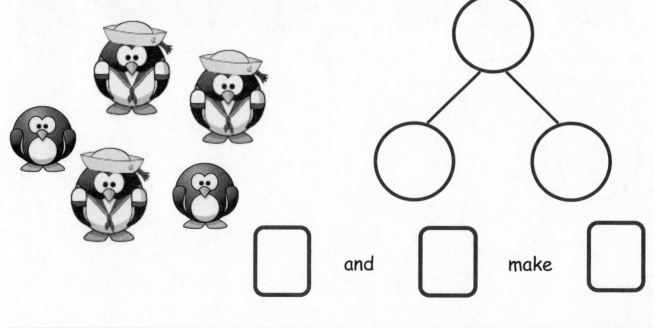

____ and ____ make ____

Lesson 5:   Represent composition and decomposition of numbers to 5 using pictorial and numeric number bonds.

©2015 Great Minds. eureka-math.org
GK-M4-SE-B3-1.3.1-01.2016

Name _____   Date _____

Fill in the number bond. Tell a story about the birds to your friend.

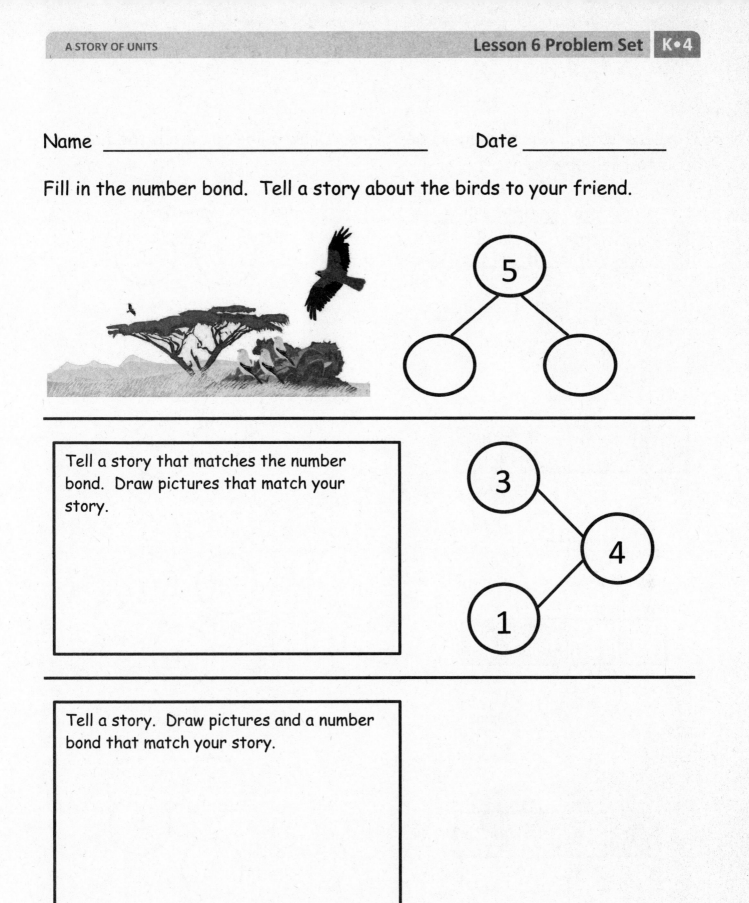

Tell a story that matches the number bond. Draw pictures that match your story.

Tell a story. Draw pictures and a number bond that match your story.

The squares below represent cube sticks. Draw a line to match the number bond to the cube stick.

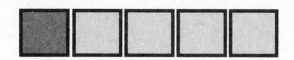

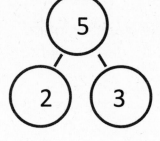

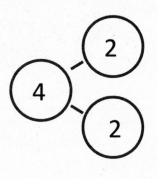

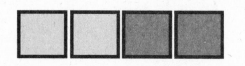

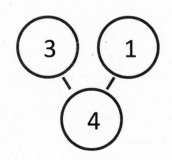

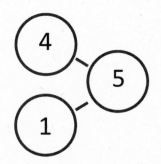

Lesson 6:   Represent number bonds with composition and decomposition story situations.

©2015 Great Minds. eureka-math.org
GK-M4-SE-B3-1.3.1-01.2016

EUREKA MATH™

Name _____ Date _____

Tell a story. Complete the number bonds. Draw pictures that match your story and number bonds

Draw some balls for your story.

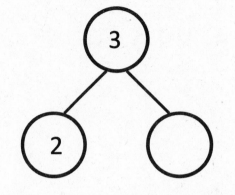

Draw some crayons for your story.

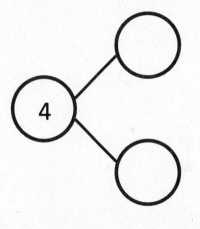

Draw some shapes for your story.

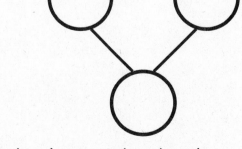

On the back of your paper, draw a picture and make a number bond.

EUREKA MATH™

Lesson 6: Represent number bonds with composition and decomposition story situations.

©2015 Great Minds. eureka-math.org
GK-M4-SE-B3-1.3.1-01.2016

21

This page intentionally left blank

Name _____ Date _____

Look at the birds. Make 2 different number bonds. Tell a friend about the numbers you put in one of the bonds.

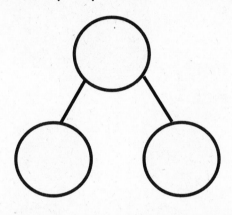 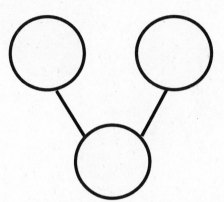

Color some squares green and the rest yellow. Write numbers in the bonds to match the colors of your squares.

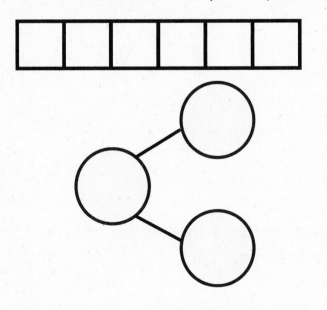 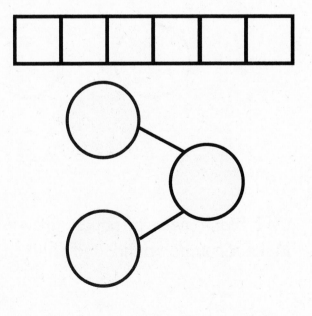

Lesson 7:   Model decompositions of 6 using a story situation, objects, and number bonds.

23

©2015 Great Minds. eureka-math.org
GK-M4-SE-B3-1.3.1-01.2016

Name _____ Date _____

Look at the presents. Make 2 different number bonds. Tell an adult about the numbers you put in the number bonds.

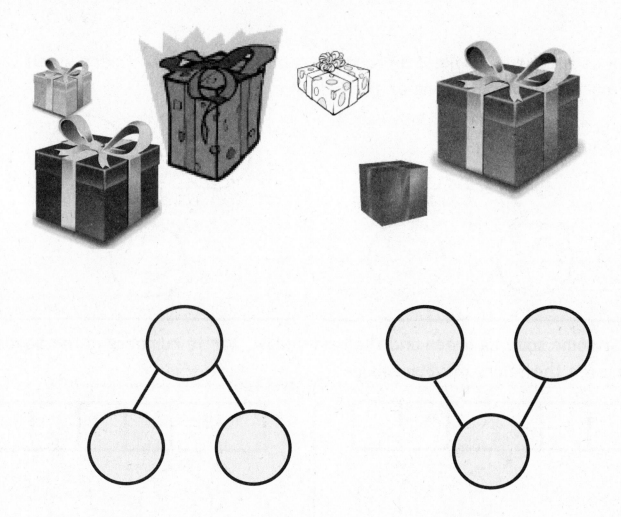

On a blank piece of paper, draw 6 presents, and sort them into 2 groups. Make a number bond, and fill it in according to your sort.

Lesson 7:   Model decompositions of 6 using a story situation, objects, and
             number bonds.

©2015 Great Minds. eureka-math.org
GK-M4-SE-B3-1.3.1-01.2016

Name _____   Date _____

Tell a story about the shapes. Complete the number bond.

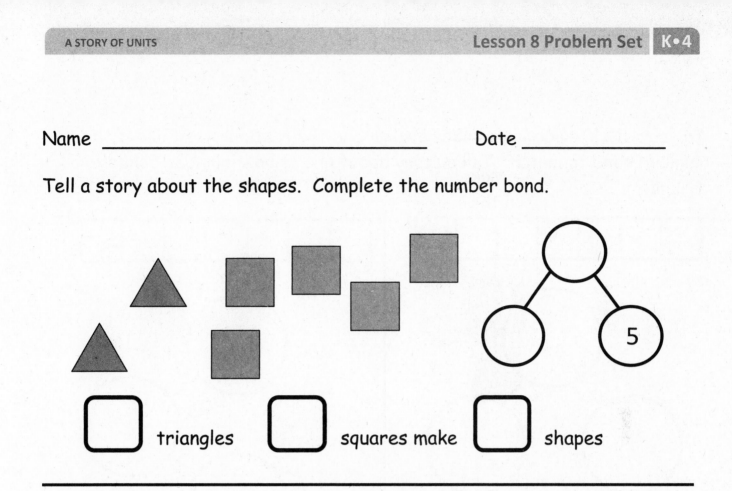

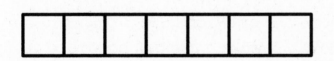 triangles   squares make   shapes

The squares below represent cube sticks. Color the cube stick to match the number bond.

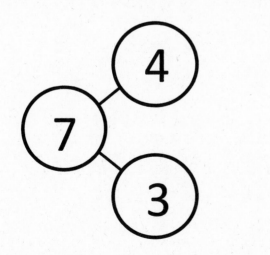

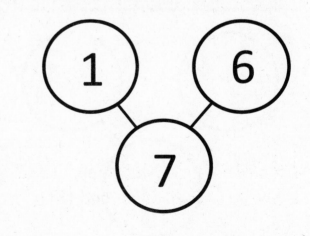

EUREKA MATH

Lesson 8:   Model decompositions of 7 using a story situation, sets, and number bonds.

25

©2015 Great Minds. eureka-math.org
GK-M4-SE-B3-1.3.1-01.2016

In each stick, color some cubes orange and the rest purple. Fill out the number bond to match. Tell a story about one of your number bonds to a friend.

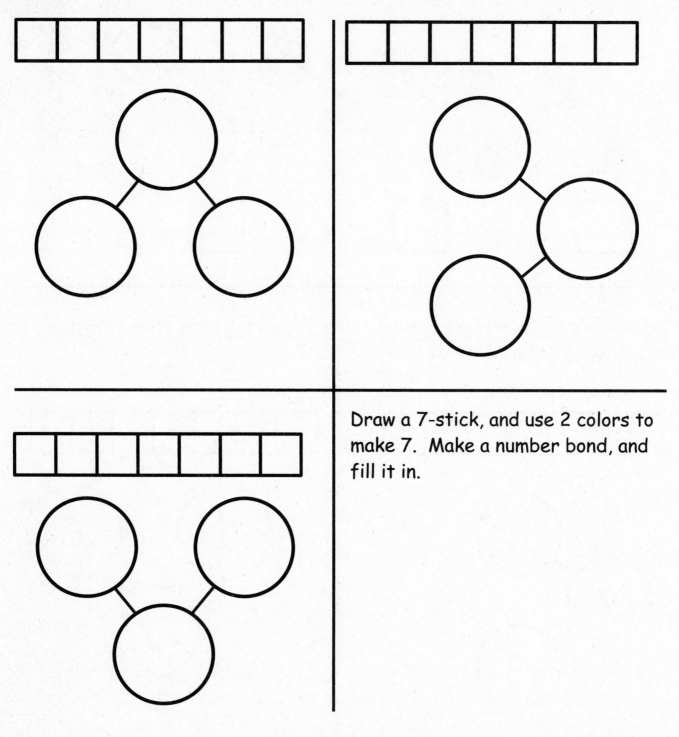

Draw a 7-stick, and use 2 colors to make 7. Make a number bond, and fill it in.

Lesson 8:   Model decompositions of 7 using a story situation, sets, and number bonds.

©2015 Great Minds. eureka-math.org
GK-M4-SE-B3-1.3.1-01.2016

Name _____   Date _____

Draw a set of 4 circles and 3 triangles. How many shapes do you have?
Fill in the number sentence and number bond.

☐   is   ☐   and   ☐

The squares represent cube sticks. Color
the cubes to match the number bond.

Color some cubes red and the rest blue.
Fill out the number bond to match.

7   5   2

On the back of your paper, draw a set of 7 squares and circles. Make a number
bond, and fill it in. Now, write a number sentence like the sentence above that
tells about your set.

Lesson 8:   Model decompositions of 7 using a story situation, sets, and
number bonds.

©2015 Great Minds. eureka-math.org
GK-M4-SE-B3-1.3.1-01.2016

27

This page intentionally left blank

Name _____   Date _____

Fill in the number bond to match the picture.

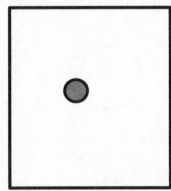

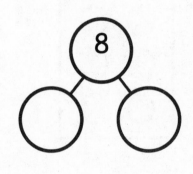

Draw some more dots to make 8 dots in all, and finish the number bond.

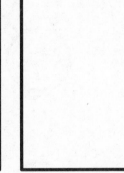

Draw 8 dots, some blue and the rest red. Fill in the number bond.

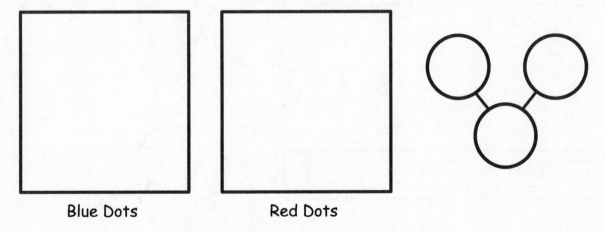

Blue Dots   Red Dots

EUREKA
MATH™

Lesson 9:   Model decompositions of 8 using a story situation, arrays, and
number bonds.

29

©2015 Great Minds. eureka-math.org
GK-M4-SE-B3-1.3.1-01.2016

Draw a line to make 2 groups of dots. Fill in the number bond.

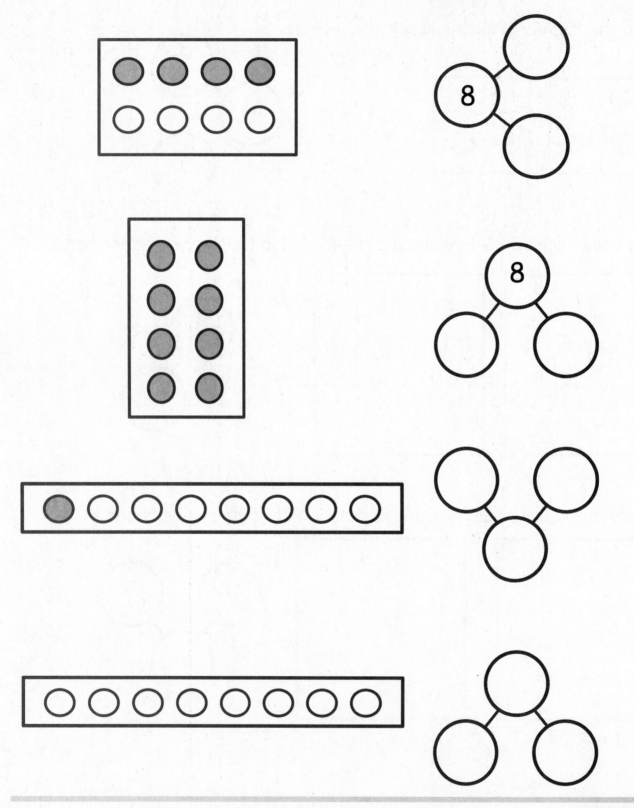

Lesson 9:   Model decompositions of 8 using a story situation, arrays, and
number bonds.

©2015 Great Minds. eureka-math.org
GK-M4-SE-B3-1.3.1-01.2016

EUREKA
MATH™

Name _____   Date _____

Complete the number bond to match the dot picture.

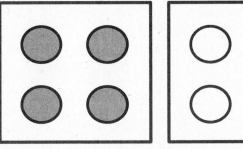 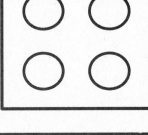 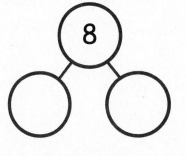

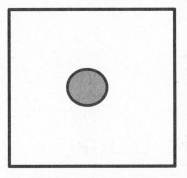 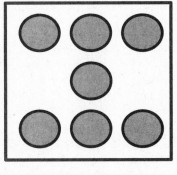 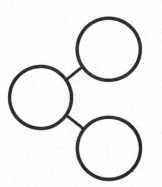

Draw a line to make 2 groups of dots. Fill in the number bond.

On the back of your paper:
- Draw a number bond for 4. Fill in the number bond.
- Draw a number bond for 5. Fill in the number bond.
- Draw a number bond for 6. Fill in the number bond.
- Draw a number bond for 7. Fill in the number bond.

This page intentionally left blank

Name _____   Date _____

Fill in the number bond to match.

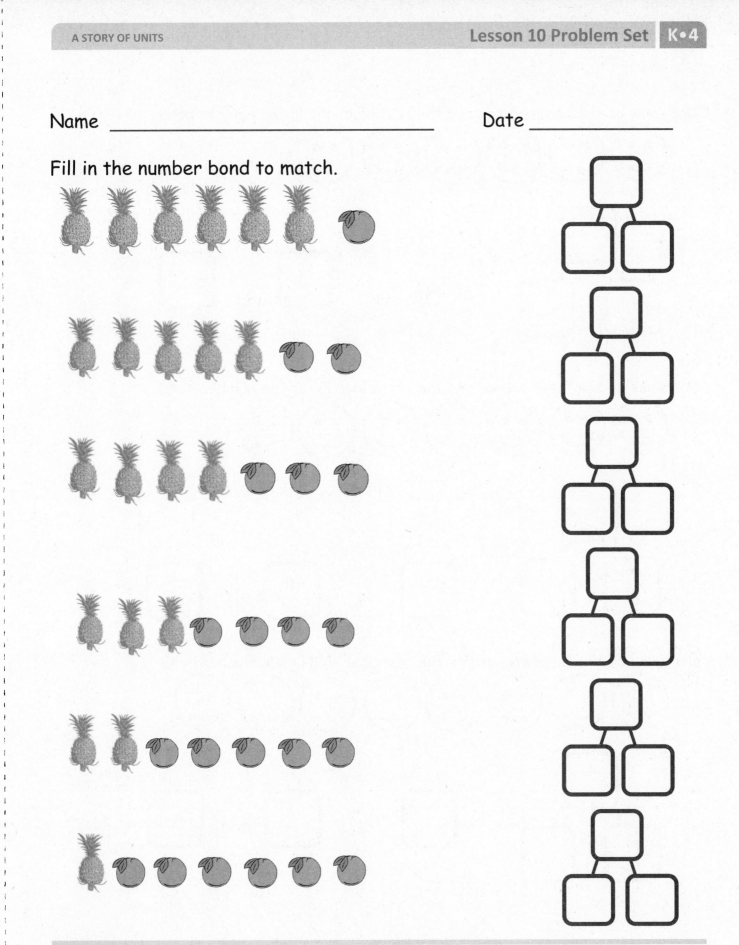

Lesson 10:   Model decompositions of 6–8 using linking cube sticks to see patterns.

©2015 Great Minds. eureka-math.org
GK-M4-SE-B3-1.3.1-01.2016

33

Color some of the faces orange and the rest blue. Fill in the number bond.

6 is ☐ and ☐

Color some of the faces orange and the rest blue. Fill in the number bond.

☐ is ☐ and ☐

Color some of the faces orange and the rest blue. Fill in the number bond.

☐ is ☐ and ☐

Lesson 10:   Model decompositions of 6–8 using linking cube sticks to see patterns.

Name _____   Date _____

These squares below represent cubes. Color 7 cubes green and 1 blue. Fill in the number bond.

[ ][ ][ ][ ][ ][ ][ ][ ]

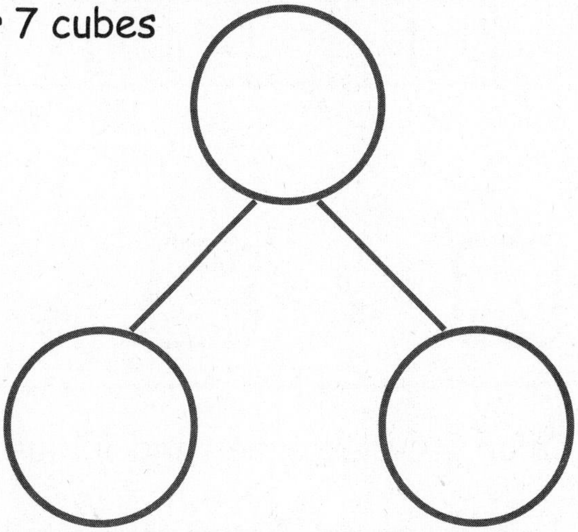

☐ is ☐ and ☐

Color 6 cubes green and 2 blue. Fill in the number bond.

[ ][ ][ ][ ][ ][ ][ ][ ]

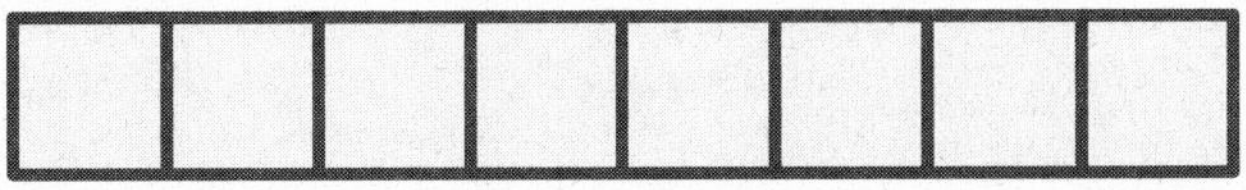

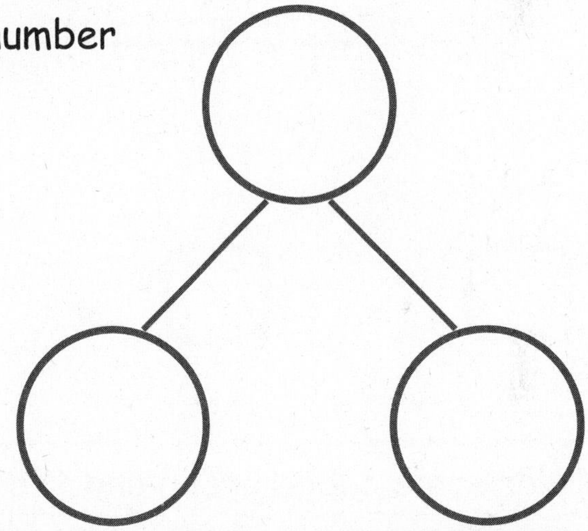

☐ is ☐ and ☐

Color some cubes green and the rest blue. Fill in the number bond.

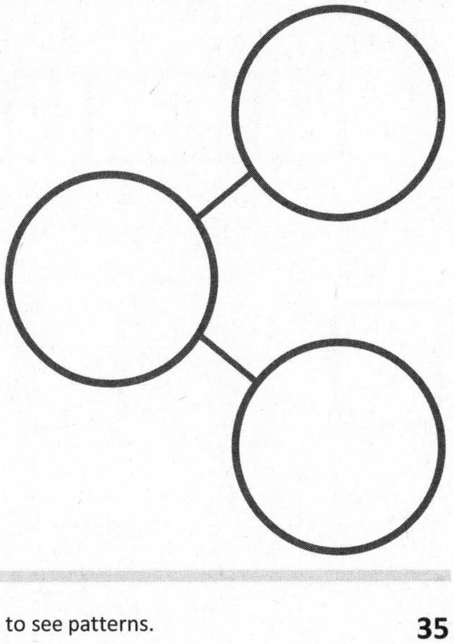

[ ][ ][ ][ ][ ][ ][ ][ ]

☐ is ☐ and ☐

Color 4 cubes green and 4 blue. Fill in the number bond.

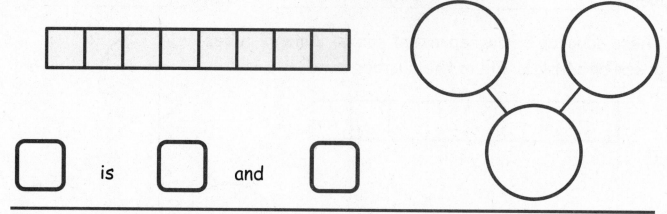

☐   is   ☐   and   ☐

Color 3 cubes green and 5 blue. Fill in the number bond.

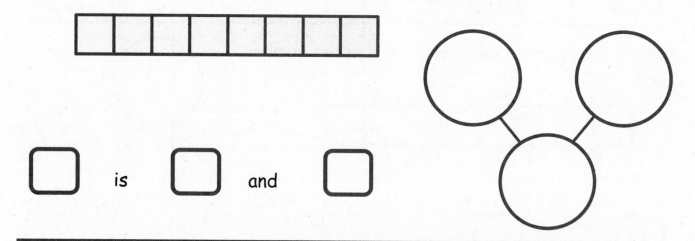

☐   is   ☐   and   ☐

Color some cubes green and the rest blue. Fill in the number bond.

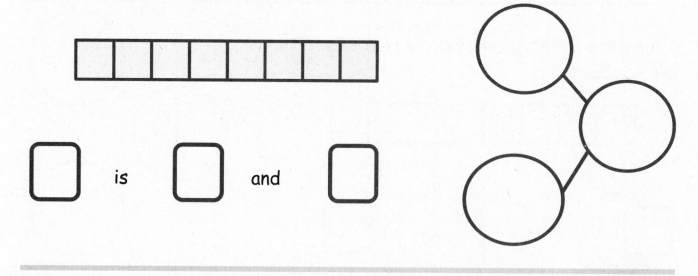

☐   is   ☐   and   ☐

Lesson 10:   Model decompositions of 6–8 using linking cube sticks to see patterns.

©2015 Great Minds. eureka-math.org
GK-M4-SE-B3-1.3.1-01.2016

Name _____ Date _____

These squares represent cubes. Draw a line to break the stick into 2 parts. Complete the number bond and number sentence.

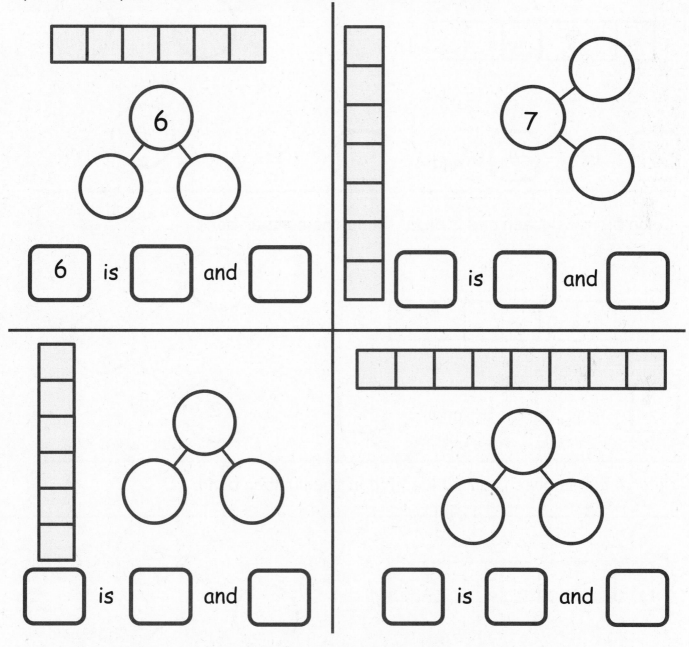

On a blank piece of paper, draw a cube stick with some red cubes and some blue cubes. Draw a number bond to match.

EUREKA MATH™

Lesson 11: Represent decompositions for 6–8 using horizontal and vertical number bonds.

©2015 Great Minds. eureka-math.org
GK-M4-SE-B3-1.3.1-01.2016

37

Name _____   Date _____

These squares represent cubes. Color 5 cubes green and 1 blue. Fill in the number bond.

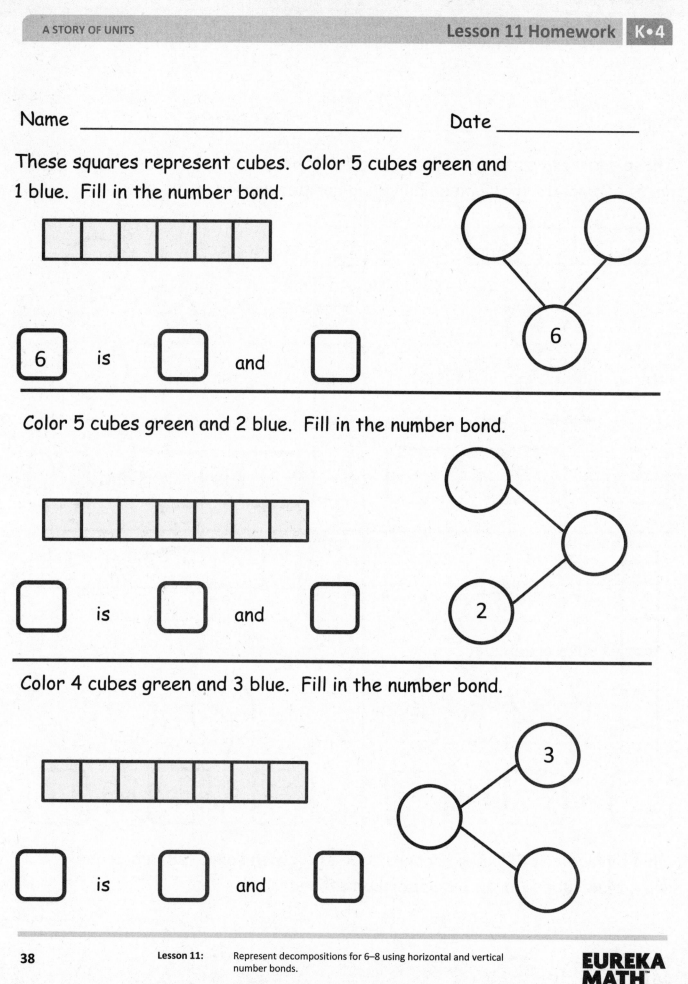

6 is ☐ and ☐

6

Color 5 cubes green and 2 blue. Fill in the number bond.

☐ is ☐ and ☐

2

Color 4 cubes green and 3 blue. Fill in the number bond.

☐ is ☐ and ☐

3

Lesson 11:   Represent decompositions for 6–8 using horizontal land vertical number bonds.

EUREKA
MATH™

Color 4 cubes green and 4 blue. Fill in the number bond.

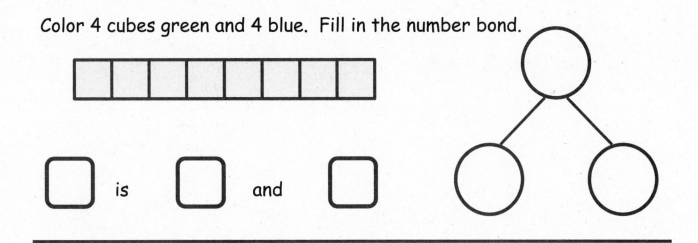

☐ is ☐ and ☐

Color 3 cubes green and 5 blue. Fill in the number bond.

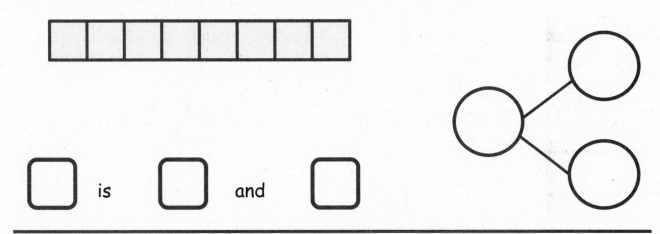

☐ is ☐ and ☐

Color 2 cubes green and 6 blue. Fill in the number bond.

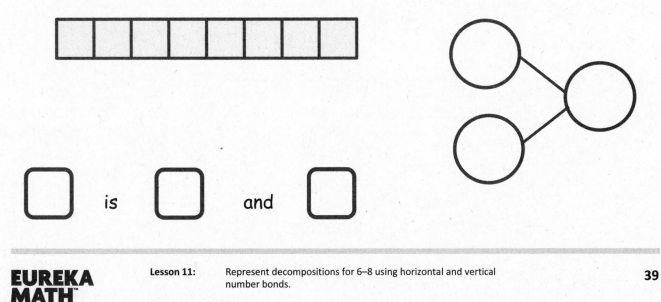

☐ is ☐ and ☐

EUREKA MATH™

Lesson 11:   Represent decompositions for 6–8 using horizontal and vertical number bonds.

©2015 Great Minds. eureka-math.org
GK-M4-SE-B3-1.3.1-01.2016

39

This page intentionally left blank

Name _____   Date _____

5 squares are colored. Color 3 more squares to make 8. Complete the number bond.

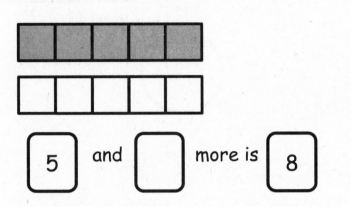

5 squares are colored. Color more squares to make 7. Complete the number bond.

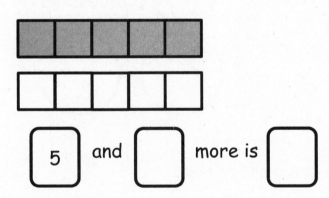
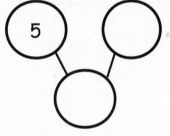

Color 6 squares. Complete the number bond.

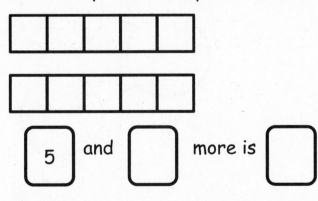
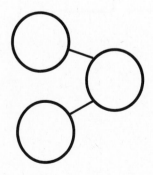

Draw more squares to make 6. Complete the number bond.

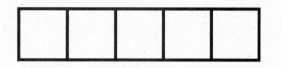

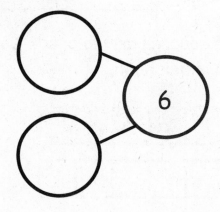

Draw more squares to make 7. Complete the number bond.

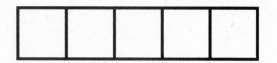

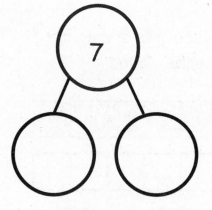

Draw more squares to make 8. Complete the number bond.

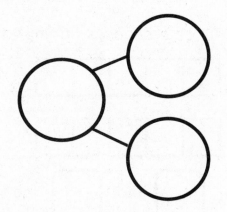

Lesson 12:   Use 5-groups to represent the 5 + n pattern to 8.

Name _____   Date _____

Fill in the number bond to match the squares.

[6] is [ ] and [1] more

6 / 1

Color 5 squares blue in the first row.

Color 2 squares red in the second row.

[ ] is [5] and [ ] more

5

Color 8 squares. Complete the number bond and sentence.

[ ] is [5] and [ ] more

5

EUREKA MATH

Lesson 12:   Use 5-groups to represent the 5 + n pattern to 8.

43

©2015 Great Minds. eureka-math.org
GK-M4-SE-B3-1.3.1-01.2016

Color the squares to match the number bond.

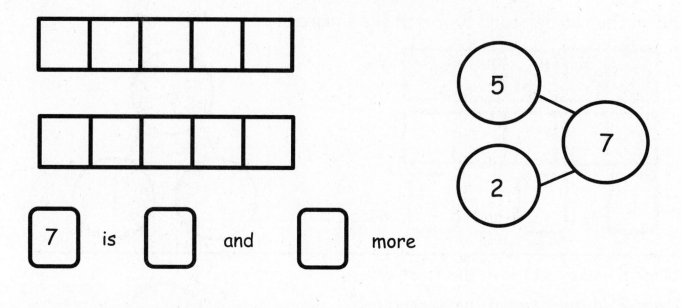

7 is ☐ and ☐ more

Color the squares to match the number bond.

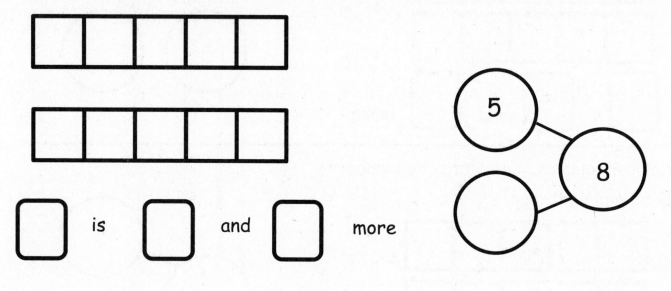

☐ is ☐ and ☐ more

Lesson 12:   Use 5-groups to represent the 5 + n pattern to 8.

©2015 Great Minds. eureka-math.org
GK-M4-SE-B3-1.3.1-01.2016

EUREKA MATH

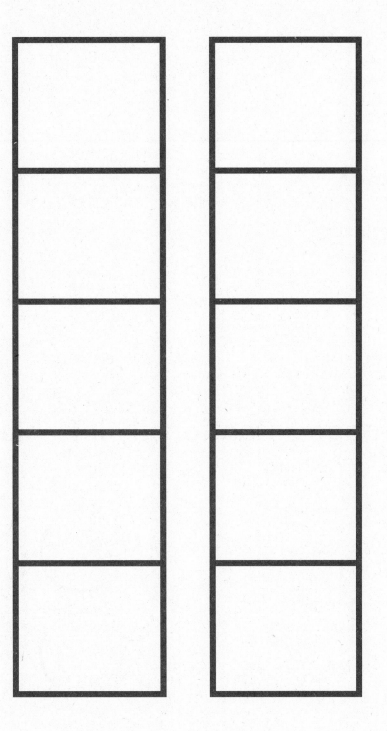

two 5-group mat

©2015 Great Minds. eureka-math.org
GK-M4-SE-B3-1.3.1-01.2016

This page intentionally left blank

Name _____   Date _____

Fill in the number bond and number sentences.

There are 6 cornstalks. 5 cornstalks are in
the first row. 1 cornstalk is in the second.

6 = ☐ + ☐

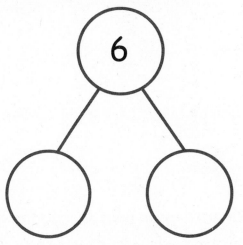

There are 6 cars on the road. 2 cars are big, and 4 are small.

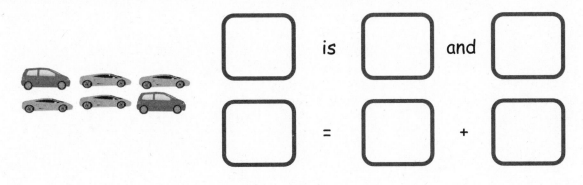

☐ is ☐ and ☐

☐ = ☐ + ☐

Lesson 13:   Represent decomposition and composition addition stories to 6 with
drawings and equations with no unknown.   47

©2015 Great Minds. eureka-math.org
GK-M4-SE-B3-1.3.1-01.2016

3 geckos have black spots, and 3 geckos have no spots. There are
6 geckos.

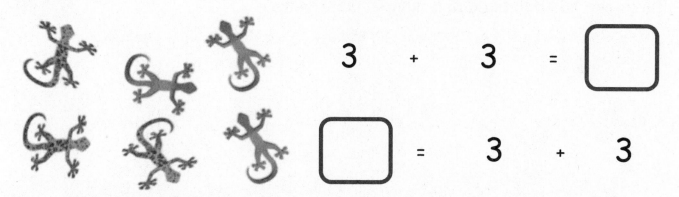

$3 \ + \ 3 \ = \ \boxed{}$

$\boxed{} \ = \ 3 \ + \ 3$

There are 6 monkeys. 4 monkeys are swinging on the tree, and 2 monkeys
are taking a nap. Draw a picture to go with the story.

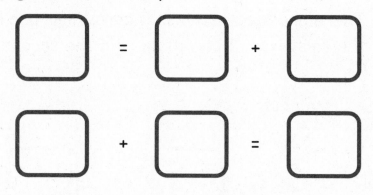

$\boxed{} \ = \ \boxed{} \ + \ \boxed{}$

$\boxed{} \ + \ \boxed{} \ = \ \boxed{}$

Create your own story, and tell your partner. Have your partner draw a
picture of your story and create a number sentence to go with the picture.

Lesson 13:   Represent decomposition and composition addition stories to 6 with
                              drawings and equations with no unknown.

Name _____   Date _____

There are 6 animals. 4 are tigers, and 2 are lions.
Fill in the number sentences and the number bond.

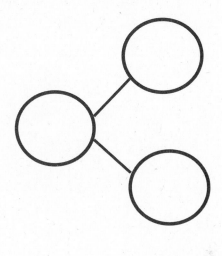

[ ] is [ ] and [ ]

[ ] = [ ] + [ ]

There 3 are monkeys and 3 elephants. All 6 animals are going into the
circus tent. Fill in the number sentences and the number bond.

[ ] and [ ] is [ ]

[ ] + [ ] = [ ]

On the back of your paper, draw some animals. Make a number bond to
match your picture.

Lesson 13: Represent decomposition and composition addition stories to 6 with
drawings and equations with no unknown.

©2015 Great Minds. eureka-math.org
GK-M4-SE-B3-1.3.1-01.2016

49

This page intentionally left blank

Name _____   Date _____

There are 7 animals. There are 5 giraffes and 2 elephants.

[   ] = 5 + 2

At the store, there was 1 big bear and 6 small bears. There were 7 bears.

1 + 6 = [   ]

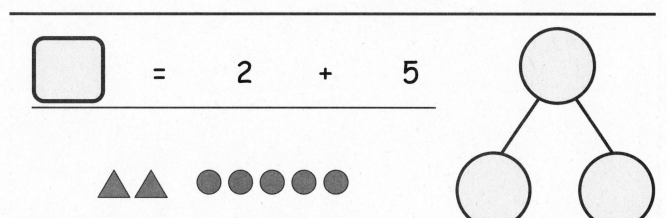

[   ] = 2 + 5

Lesson 14:   Represent decomposition and composition addition stories to 7 with
drawings and equations with no unknown.

©2015 Great Minds. eureka-math.org
GK-M4-SE-B3-1.3.1-01.2016

51

The squares below represent cubes.
4 gray cubes and 3 white cubes are 7 cubes.

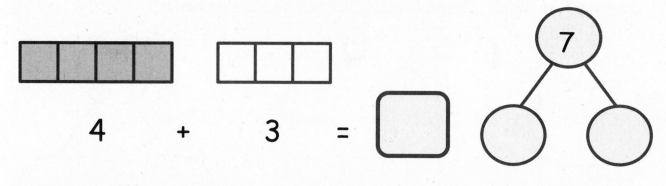

4   +   3   =

Color the cubes to match the cubes above. Fill in the number sentence.

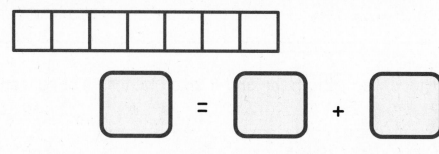

☐ = ☐ + ☐

Create your own story, and tell your partner. Have your partner draw
a picture of your story and create a number sentence to go with the
picture.

Lesson 14:   Represent decomposition and composition addition stories to 7 with
                         drawings and equations with no unknown.

©2015 Great Minds. eureka-math.org
GK-M4-SE-B3-1.3.1-01.2016

Name _____   Date _____

There are 7 bears. 3 bears have bowties. 4 bears have hearts. Fill in the number sentences and the number bond.

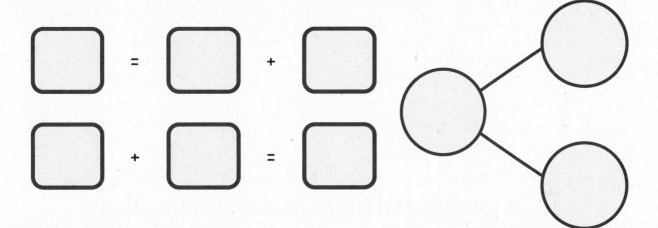

5 bears have scarves on, and 2 do not. There are 7 bears.
Write a number sentence that tells about the bears.

On the back of your paper, draw a picture about the 7 bears. Write a number sentence, and make a number bond to go with it.

Lesson 14:   Represent decomposition and composition addition stories to 7 with
drawings and equations with no unknown.

©2015 Great Minds. eureka-math.org
GK-M4-SE-B3-1.3.1-01.2016

53

This page intentionally left blank

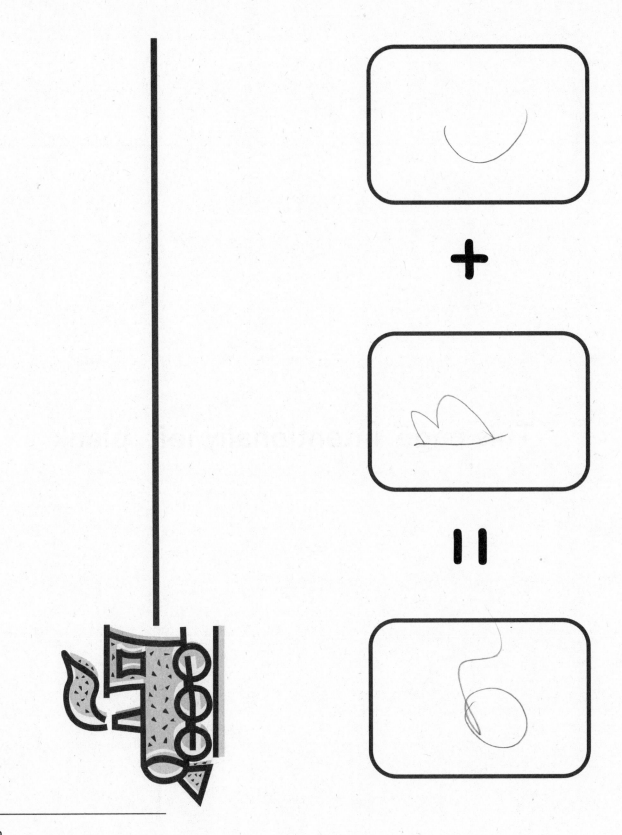

train

Lesson 14:   Represent decomposition and composition addition stories to 7 with
drawings and equations with no unknown.

55

©2015 Great Minds. eureka-math.org
GK-M4-SE-B3-1.3.1-01.2016

This page intentionally left blank

Name LIACHATTER   Date _____

Fill in the number sentences.

There are 8 fish. There are 4 striped fish and 4 goldfish.

$8 = 4 + 4$

$4 + 4 = 8$

There are 8 shapes. There are 5 triangles and 3 diamonds.

$8 = 5 + 2$

$5 + 3 = 8$

There are 6 stars and 2 moons.
There are 8 shapes.

$2 + 6 = 8$

$8 = 6 + 2$

Lesson 15:   Represent decomposition and composition addition stories to 8 with
drawings and equations with no unknown.

57

©2015 Great Minds. eureka-math.org
GK-M4-SE-B3-1.3.1-01.2016

There are 8 shapes. Count and circle the squares. Count and circle the triangle.

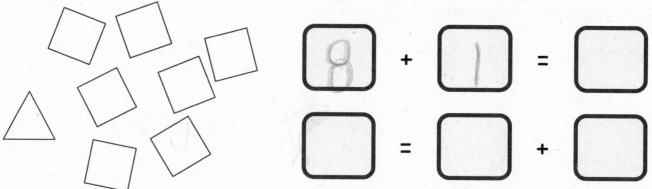

There are 8 flowers. Some flowers are yellow, and some flowers are red. Draw a picture to go with the story.

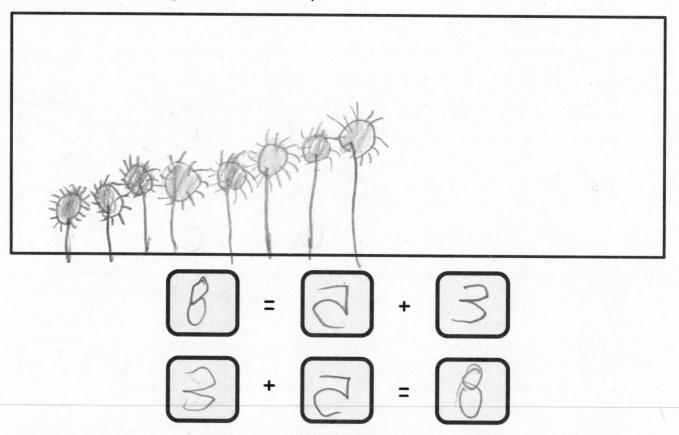

Create your own story, and tell your partner. Have your partner draw a picture of your story and create a number sentence to go with the picture.

Lesson 15:   Represent decomposition and composition addition stories to 8 with drawings and equations with no unknown.

©2015 Great Minds. eureka-math.org
GK-M4-SE-B3-1.3.1-01.2016

Name _____   Date _____

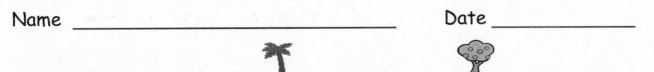

There are 8 trees. 5 are palm trees, and 3 are apple trees. Fill in the number sentences and the number bond.

⬚ = ⬚ + ⬚

⬚ + ⬚ = ⬚

There are 8 trees. 4 are oak trees, and 4 are spruce trees. Fill in the number sentences and the number bond.

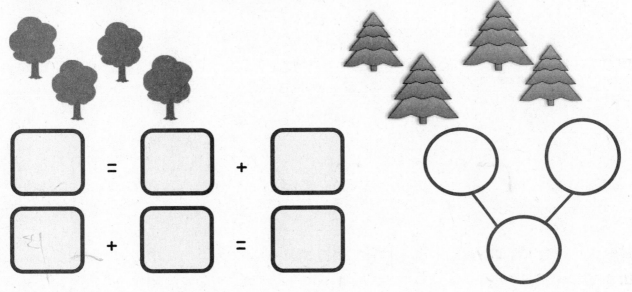

⬚ = ⬚ + ⬚

⬚ + ⬚ = ⬚

Lesson 15:   Represent decomposition and composition addition stories to 8 with drawings and equations with no unknown.   59

©2015 Great Minds. eureka-math.org
GK-M4-SE-B3-1.3.1-01.2016

This page intentionally left blank

Name _____   Date _____

There are 4 snakes sitting on the rocks. 2 more snakes slither over. How many snakes are on the rocks now? Put a box around all the snakes, trace the mystery box, and write the answer inside it.

4   +   2   =

There are 5 turtles swimming. Draw 2 more turtles that come to swim. How many turtles are swimming now? Draw a box around all the turtles, draw a mystery box, and write the answer.

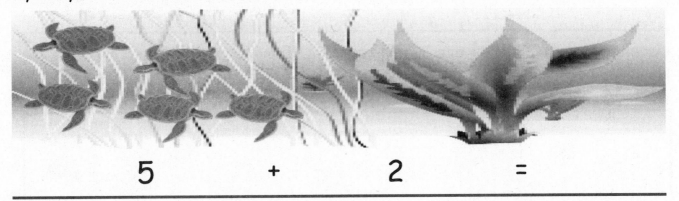

5   +   2   =

Today is your birthday! You have 7 presents. A friend brings another present. Draw the present. How many presents are there now? Draw a mystery box, and write the answer inside it.

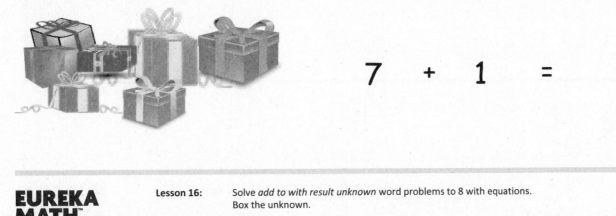

7   +   1   =

Listen and draw. There were 6 girls playing soccer. A boy came to play. How many children were playing soccer then? Draw a box around all the children.

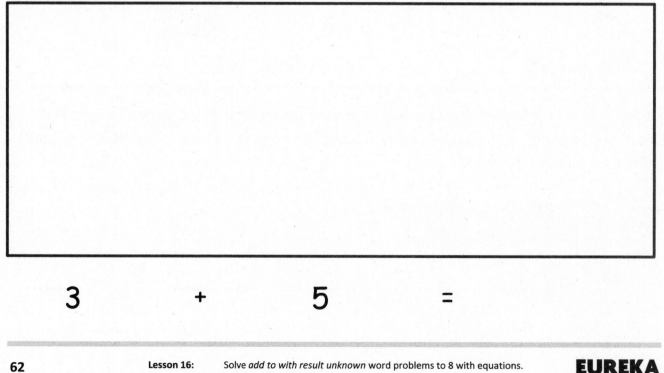

6   +   1   =   [   ]

Listen and draw. There were 3 frogs on a log. 5 more frogs hopped onto the log. How many frogs were on the log then? Draw a box around the frogs, and box the answer.

3   +   5   =

Lesson 16:   Solve *add to with result unknown* word problems to 8 with equations.
                           Box the unknown.

©2015 Great Minds. eureka-math.org
GK-M4-SE-B3-1.3.1-01.2016

EUREKA
MATH

Name _____   Date _____

There are 3 penguins on the ice.
4 more penguins are coming.
How many penguins are there?

3   +   4   =   [ ]

There is 1 mama bear. 5 baby
bears are following her. How many
bears are there? Draw a box for
the answer.

1   +   5   =

Draw 7 balls in the ball box. Draw a girl putting 1 more ball in the ball
box. Circle all the balls, and draw a box for the answer. Write your
answer.

7   +   1   =

EUREKA MATH

Lesson 16: Solve *add to with result unknown* word problems to 8 with equations.
Box the unknown.

©2015 Great Minds. eureka-math.org
GK-M4-SE-B3-1.3.1-01.2016

63

This page intentionally left blank

Name _____   Date _____

There are 4 green balloons and 3 orange balloons in the air. How many balloons are in the air? Color the balloons to match the story, and fill in the number sentences.

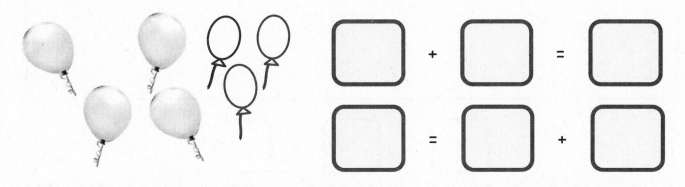

☐ + ☐ = ☐

☐ = ☐ + ☐

Dominic has 6 yellow star stickers and 2 blue star stickers. How many stickers does Dominic have? Color the stars to match the story, and fill in the number sentences.

☐ + ☐ = ☐

☐ = ☐ + ☐

There are 5 big robots and 1 little robot. How many robots are there? Fill in the number sentences.

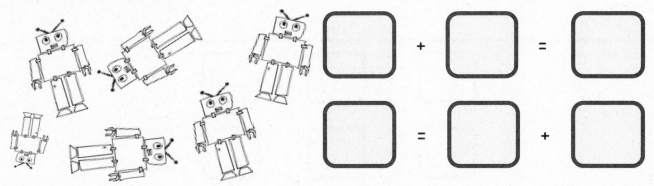

☐ + ☐ = ☐

☐ = ☐ + ☐

Lesson 17:   Solve *put together with total unknown* word problems to 8 using objects and drawings.

©2015 Great Minds. eureka-math.org
GK-M4-SE-B3-1.3.1-01.2016

65

Listen and draw. Charlotte is playing with pattern blocks. She has 3 squares and 3 triangles. How many shapes does Charlotte have?

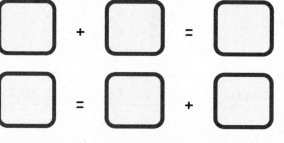

Listen and draw. Gavin is making a tower with linking cubes. He has 5 purple and 3 orange cubes. How many linking cubes does Gavin have?

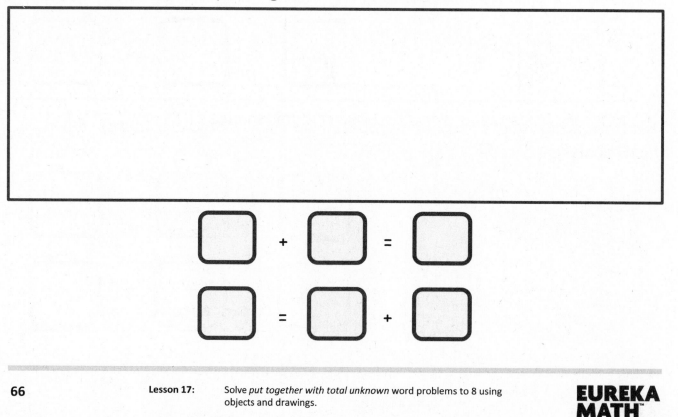

Lesson 17:   Solve *put together with total unknown* word problems to 8 using objects and drawings.

EUREKA MATH™

Name _____   Date _____

There are 5 hexagons and 2 triangles. How many shapes are there?

☐ = ☐ + ☐

☐ + ☐ = ☐

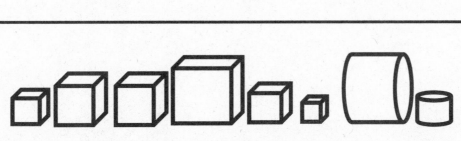

There are 6 cubes and 2 cylinders. How many shapes are there?

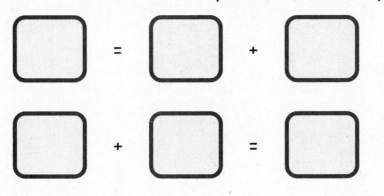

☐ = ☐ + ☐

☐ + ☐ = ☐

On the back of your paper, draw some shapes, and make a number sentence to match.

Lesson 17:   Solve *put together with total unknown* word problems to 8 using objects and drawings.

67

©2015 Great Minds. eureka-math.org
GK-M4-SE-B3-1.3.1-01.2016

This page intentionally left blank

tree and sun

Lesson 17:   Solve *put together with total unknown* word problems to 8 using
objects and drawings.

69

©2015 Great Minds. eureka-math.org
GK-M4-SE-B3-1.3.1-01.2016

This page intentionally left blank

Name _____   Date _____

Devin has 6 Spiderman pencils. He put some in his desk and the rest in his pencil box. Write a number sentence to show how many pencils Devin might have in his desk and pencil box.

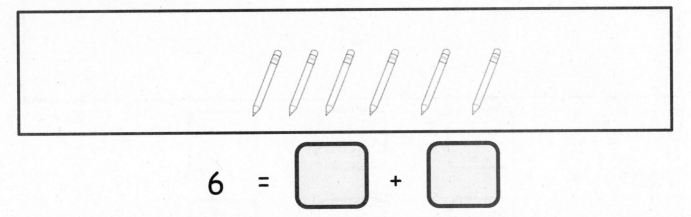

6 = [   ] + [   ]

Shania made 7 necklaces. She wore some of the necklaces and put the rest in her jewelry box. Use the linking cubes to help you think about how many necklaces Shania might have on and how many are in her jewelry box. Then, complete the number sentences.

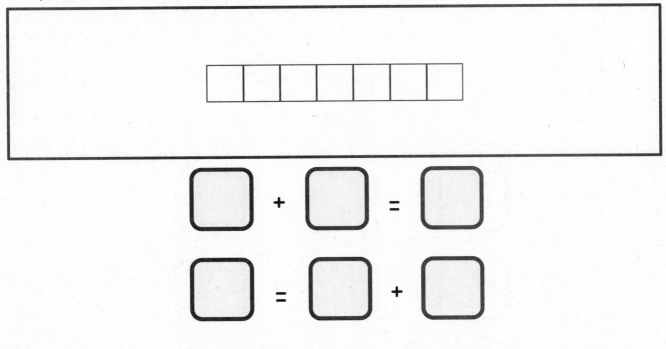

[   ] + [   ] = [   ]

[   ] = [   ] + [   ]

Tommy planted 8 flowers. He planted some in his garden and some in flowerpots.
Draw how Tommy may have planted the flowers. Fill in the number sentences to
match your picture.

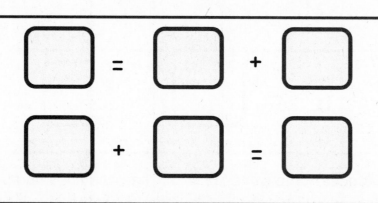

Create your own story, and draw a picture. Fill in the number sentences.
Tell your story to a friend.

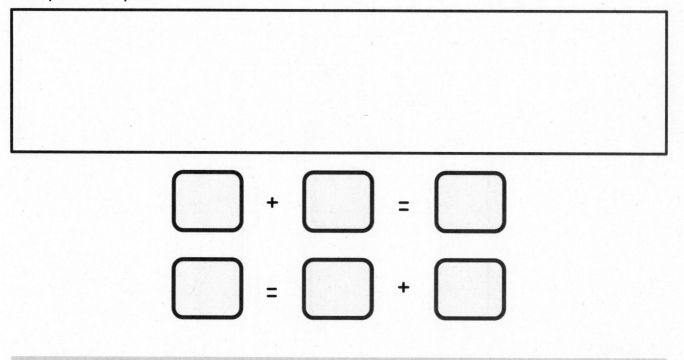

Lesson 18:   Solve *both addends unknown* word problems to 8 to find addition
                       patterns in number pairs.

©2015 Great Minds. eureka-math.org
GK-M4-SE-B3-1.3.1-01.2016

Name _____   Date _____

Ted has 7 toy cars. Color some cars red and the rest blue. Write a number sentence that shows how many are red and how many are blue.

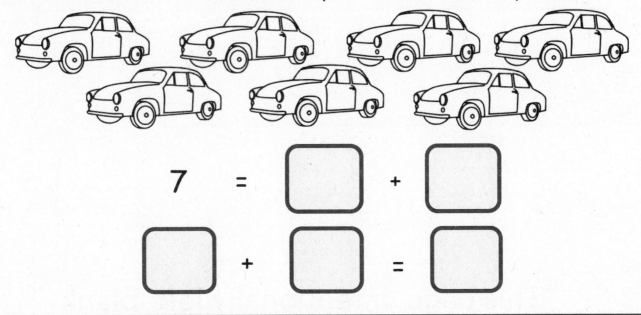

7 = ☐ + ☐

☐ + ☐ = ☐

Chuck has 8 balls. Some are red, and the rest are blue. Color to show Chuck's balls. Fill in the number sentences.

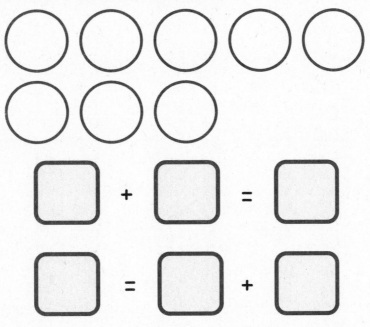

☐ + ☐ = ☐

☐ = ☐ + ☐

Lesson 18:   Solve *both addends unknown* word problems to 8 to find addition patterns in number pairs.

©2015 Great Minds. eureka-math.org
GK-M4-SE-B3-1.3.1-01.2016

73

This page intentionally left blank

Name _____   Date _____

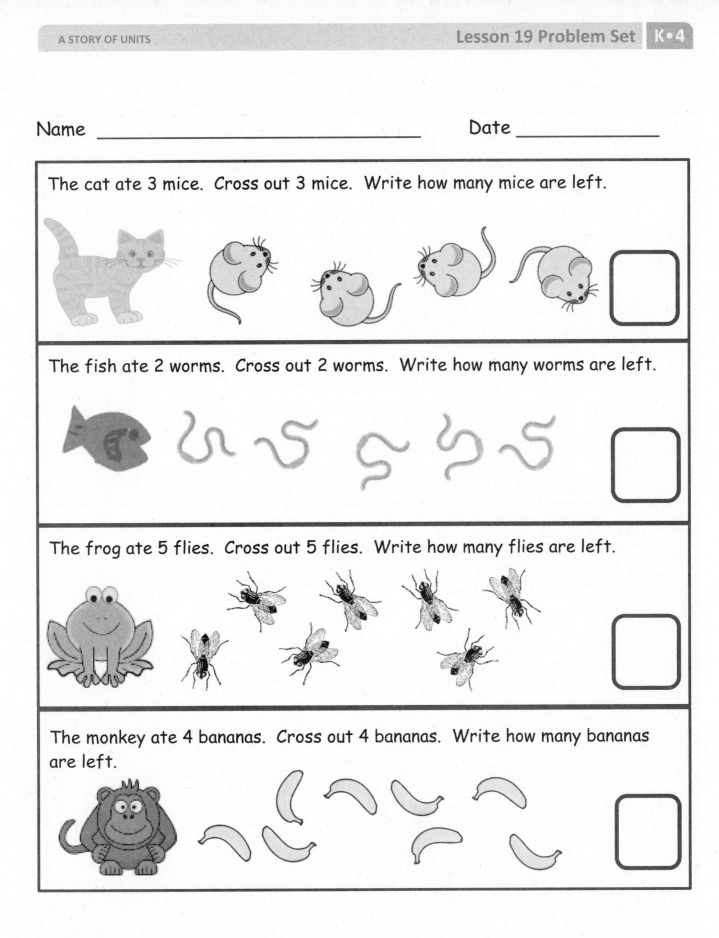

The cat ate 3 mice. Cross out 3 mice. Write how many mice are left.

The fish ate 2 worms. Cross out 2 worms. Write how many worms are left.

The frog ate 5 flies. Cross out 5 flies. Write how many flies are left.

The monkey ate 4 bananas. Cross out 4 bananas. Write how many bananas are left.

Draw 6 balls. The boy kicked 3 balls down the hill. How many balls does he have left?

There are 5 butterflies flying around the flower. Draw them. 1 of the butterflies flew away, so cross it out. How many butterflies are left?

Lesson 19:   Use objects and drawings to find *how many are left.*

Name _____ Date _____

1 train drove away. Cross out 1. Write how many were left.

2 horses were bought. Cross out 2. How many were left at the store?

4 ducks swam away. Cross out 4. Write how many are left.

There are 7 apples in the tree. Draw them. A bird ate 1 of them, so cross it out. How many apples are left?

EUREKA
MATH™

Lesson 19:   Use objects and drawings to find *how many are left*.

©2015 Great Minds. eureka-math.org
GK-M4-SE-B3-1.3.1-01.2016

77

This page intentionally left blank

Name _____   Date _____

Draw a line from the picture to the number sentence it matches.

3 – 1 = 2

5 – 4 = 1

4 – 2 = 2

5 – 1 = 4

Pick 1 mouse picture, and tell a story to your partner. See if your partner can pick the picture you told the story about.

Lesson 20:   Solve *take from with result unknown* expressions and equations using the minus sign with no unknown.

©2015 Great Minds. eureka-math.org
GK-M4-SE-B3-1.3.1-01.2016

79

Cross out the bears to match the number sentences.

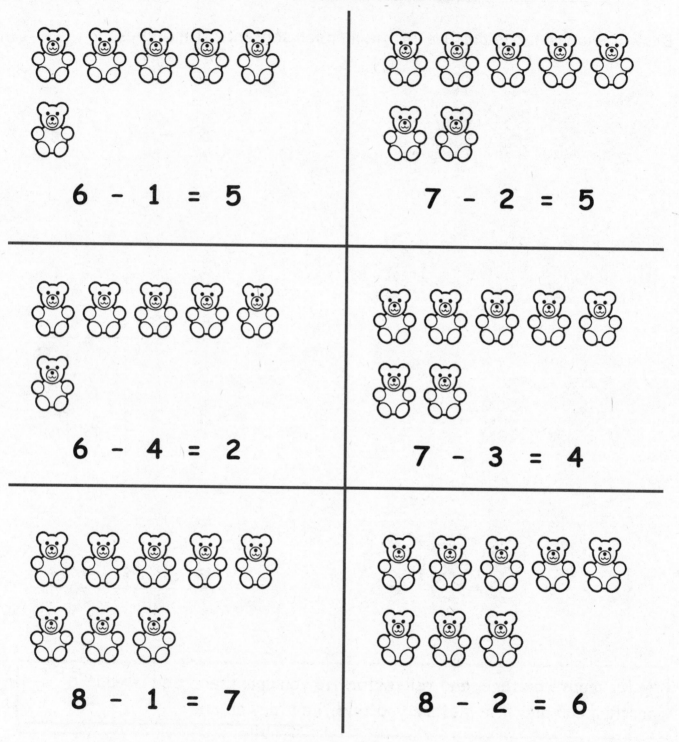

6 – 1 = 5

7 – 2 = 5

6 – 4 = 2

7 – 3 = 4

8 – 1 = 7

8 – 2 = 6

Lesson 20:   Solve *take from with result unknown* expressions and equations using the minus sign with no unknown.

©2015 Great Minds. eureka-math.org
GK-M4-SE-B3-1.3.1-01.2016

Name _____   Date _____

The squares below represent cube sticks. Match the cube stick to the number sentence.

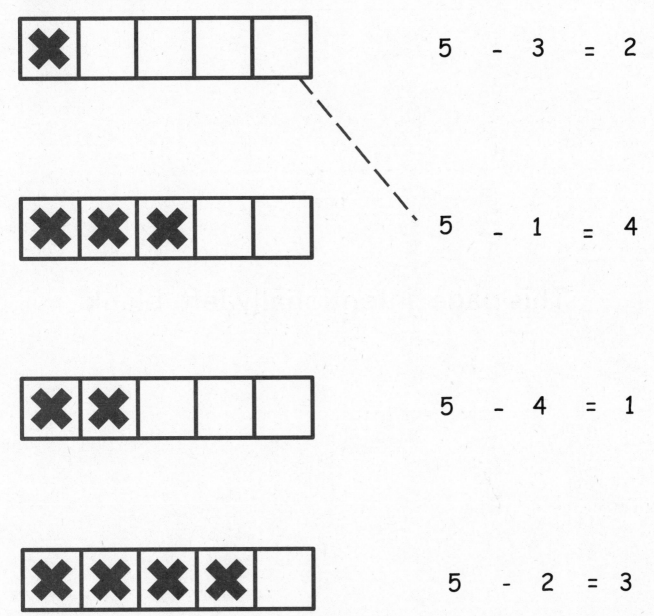

5 - 3 = 2

5 - 1 = 4

5 - 4 = 1

5 - 2 = 3

On the back of the paper, draw a 5-stick, cross out some cubes, and write a number sentence.

Lesson 20:   Solve *take from with result unknown* expressions and equations using the minus sign with no unknown.

©2015 Great Minds. eureka-math.org
GK-M4-SE-B3-1.3.1-01.2016

81

This page intentionally left blank

Name _____   Date _____

Tyler bought a cone with 4 scoops. He ate 1 scoop. Cross out 1 scoop. How many scoops were left?

4   -   1   =   ☐

Eva ate ice cream, too. She ate 2 scoops. How many scoops were left?

4   -   2   =   ☐

There were 4 bottles. 3 of them broke. How many bottles were left?

4   -   3   =   ☐

Lesson 21:   Represent subtraction story problems using objects, drawings, expressions, and equations.

83

©2015 Great Minds. eureka-math.org
GK-M4-SE-B3-1.3.1-01.2016

Anthony had 5 erasers in his pencil box. He dropped his pencil box, and 4 erasers fell on the floor. How many erasers are in Anthony's pencil box now? Draw the erasers, and fill in the number sentence.

5 - 4 = ☐

Tanisha had 5 grapes. She gave 3 grapes to a friend. How many grapes does Tanisha have now? Draw the grapes, and fill in the number sentence.

☐ - ☐ = ☐

Lesson 21:   Represent subtraction story problems using objects, drawings, expressions, and equations.

©2015 Great Minds. eureka-math.org
GK-M4-SE-B3-1.3.1-01.2016

EUREKA MATH

Name _____   Date _____

There were 5 apples. Bill ate 1. Cross out the apple he ate. How many apples were left? Fill in the boxes.

5 take away 1 is [   ]

5 – 1 = [   ]

There were 5 oranges. Pat took 2. Draw the oranges. Cross out the 2 she took. How many oranges were left? Fill in the boxes.

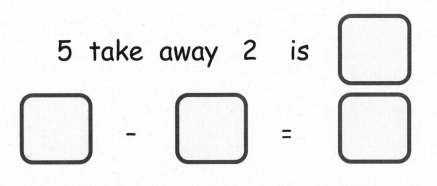

5 take away 2 is [   ]

[   ] – [   ] = [   ]

This page intentionally left blank

Name _____   Date _____

Fill in the number bonds.

Cross out 1 hat.

6 – 1 = 5

Cross out 5 snowflakes.

6 – 5 = 1

Cross out 2 snowflakes.

6 – 2 = 4

Lesson 22:   Decompose the number 6 using 5-group drawings by breaking off or
removing a part, and record each decomposition with a drawing and
subtraction equation.

©2015 Great Minds. eureka-math.org
GK-M4-SE-B3-1.3.1-01.2016

87

Fill in the number sentences and the number bonds.

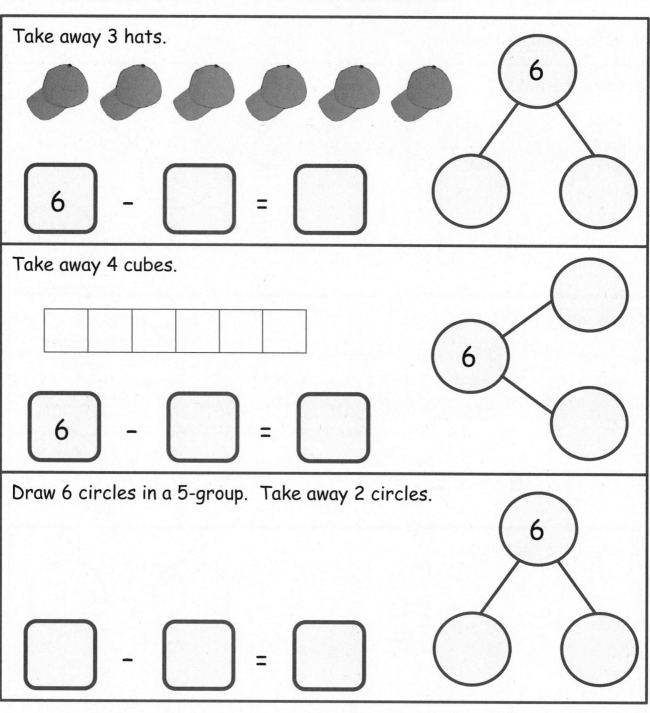

Take away 3 hats.

6 − ☐ = ☐

6

Take away 4 cubes.

6 − ☐ = ☐

6

Draw 6 circles in a 5-group. Take away 2 circles.

☐ − ☐ = ☐

6

Lesson 22:   Decompose the number 6 using 5-group drawings by breaking off or
removing a part, and record each decomposition with a drawing and
subtraction equation.

©2015 Great Minds. eureka-math.org
GK-M4-SE-B3-1.3.1-01.2016

EUREKA
MATH™

Name _____   Date _____

Here are 6 books. Cross out 2. How many are left? Fill in the number bond and the number sentence.

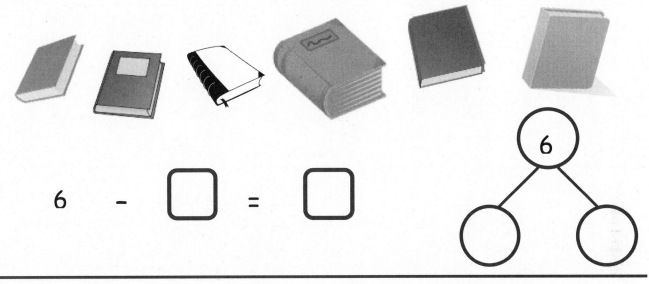

6 - ☐ = ☐

Draw 6 stars. Cross out 4. Fill in the number sentence and the number bond.

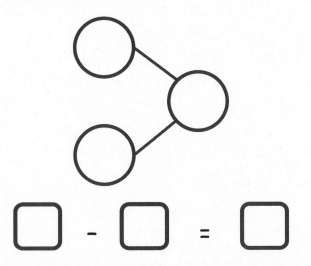

☐ - ☐ = ☐

Draw 6 objects. Cross out 5. Fill in the number sentence and the number bond.

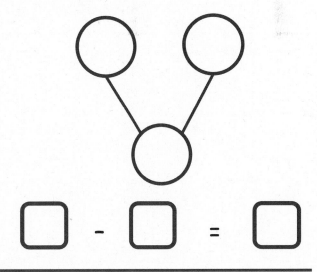

☐ - ☐ = ☐

On the back of your paper, draw 6 triangles. Cross out 1. Write a number sentence, and draw a number bond to match.

Lesson 22:   Decompose the number 6 using 5-group drawings by breaking off or removing a part, and record each decomposition with a drawing and subtraction equation.

©2015 Great Minds. eureka-math.org
GK-M4-SE-B3-1.3.1-01.2016

89

This page intentionally left blank

Name _____   Date _____

Say the number sentence. Fill in the blanks. Cross out the number.
Cross out 2 dots.

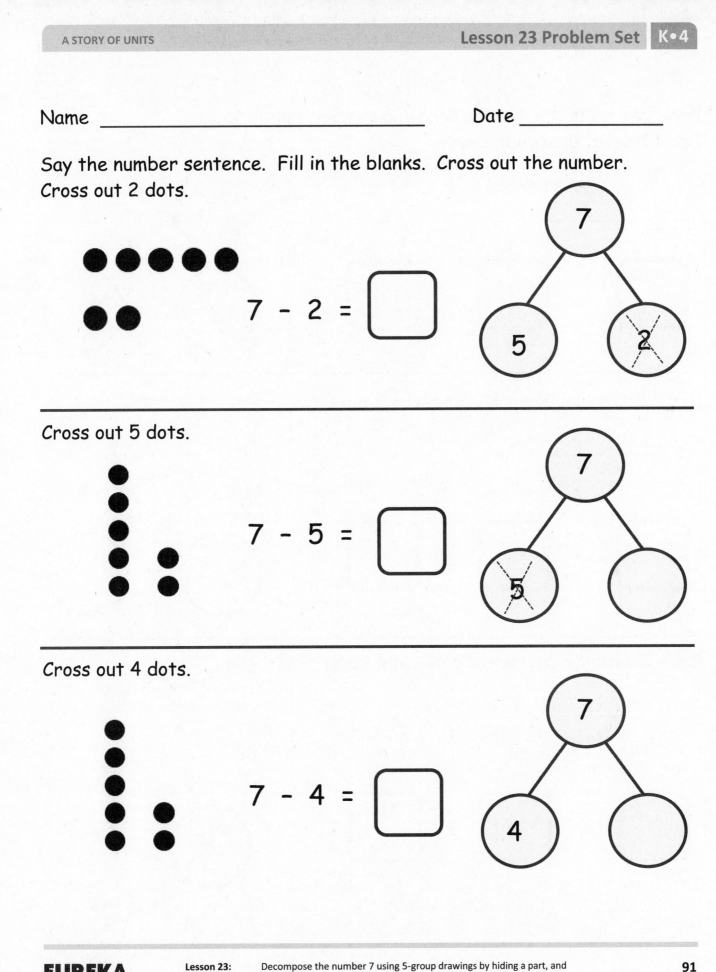

7 – 2 = ☐

Cross out 5 dots.

7 – 5 = ☐

Cross out 4 dots.

7 – 4 = ☐

EUREKA MATH   **Lesson 23:**   Decompose the number 7 using 5-group drawings by hiding a part, and   **91**
                              record each decomposition with a drawing and subtraction equation.

©2015 Great Minds. eureka-math.org
GK-M4-SE-B3-1.3.1-01.2016

Draw and fill in the number bond and number sentence.
Draw 7 dots. Cross out 2 dots.

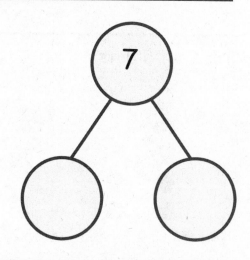

Draw 7 dots in a 5-group. Cross out 3 dots.

Draw 7 dots in a 5-group. Cross out 4 dots.

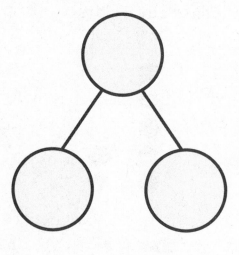

Lesson 23: Decompose the number 7 using 5-group drawings by hiding a part, and record each decomposition with a drawing and subtraction equation.

Name _____   Date _____

Fill in the number sentence and number bond.
Cross out 5 dots.

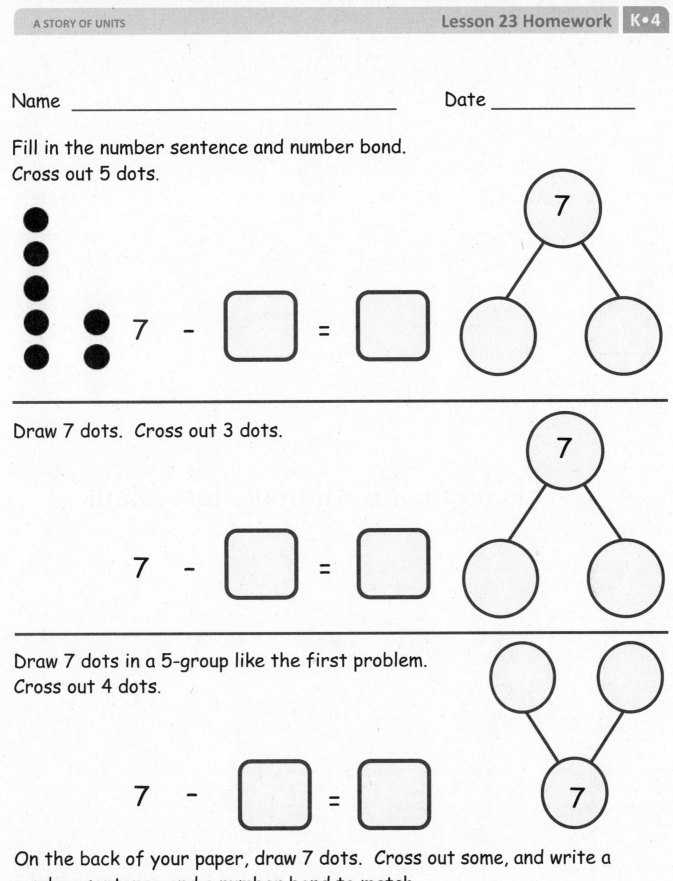

7 − ☐ = ☐

7

Draw 7 dots. Cross out 3 dots.

7 − ☐ = ☐

7

Draw 7 dots in a 5-group like the first problem.
Cross out 4 dots.

7 − ☐ = ☐

7

On the back of your paper, draw 7 dots. Cross out some, and write a
number sentence and a number bond to match.

Lesson 23:   Decompose the number 7 using 5-group drawings by hiding a part, and
record each decomposition with a drawing and subtraction equation.

93

©2015 Great Minds. eureka-math.org
GK-M4-SE-B3-1.3.1-01.2016

This page intentionally left blank

Name _____   Date _____

Fill in the number sentences and number bonds.

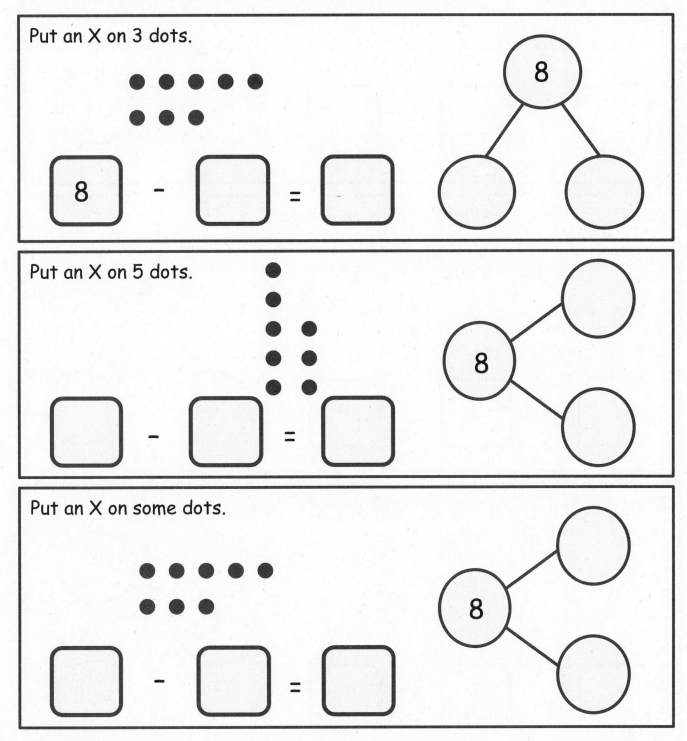

Put an X on 3 dots.

8 − ☐ = ☐

8

Put an X on 5 dots.

☐ − ☐ = ☐

8

Put an X on some dots.

☐ − ☐ = ☐

8

Lesson 24:   Decompose the number 8 using 5-group drawings and crossing off a
part, and record each decomposition with a drawing and subtraction
equation.

©2015 Great Minds. eureka-math.org
GK-M4-SE-B3-1.3.1-01.2016

95

Draw 8 dots. Put an X on 1 dot.

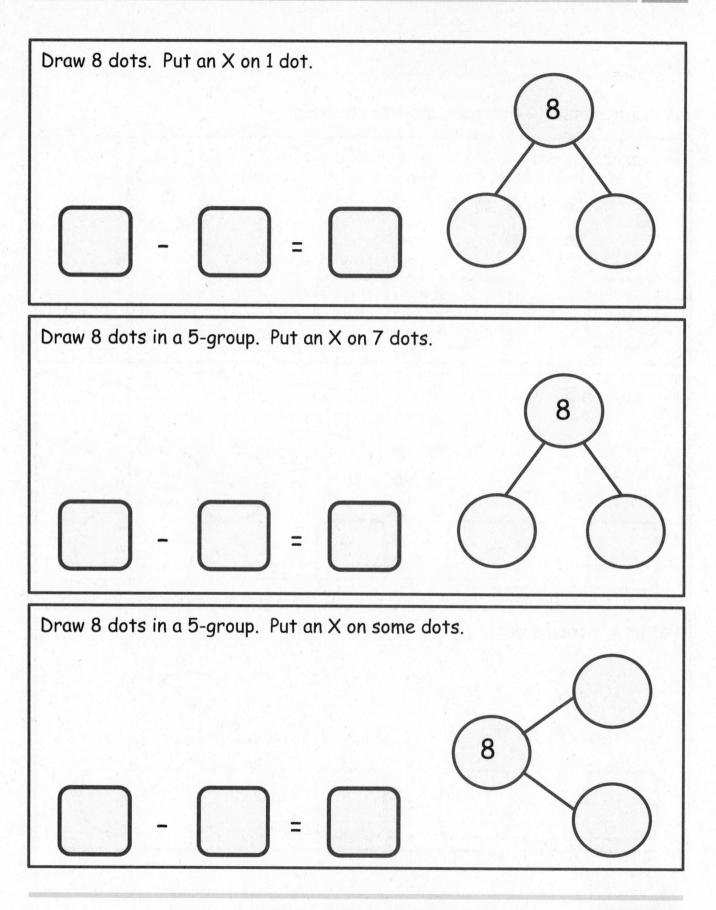

◻ - ◻ = ◻

Draw 8 dots in a 5-group. Put an X on 7 dots.

◻ - ◻ = ◻

Draw 8 dots in a 5-group. Put an X on some dots.

◻ - ◻ = ◻

Lesson 24:   Decompose the number 8 using 5-group drawings and crossing off a part, and record each decomposition with a drawing and subtraction equation.

©2015 Great Minds. eureka-math.org
GK-M4-SE-B3-1.3.1-01.2016

Name _____   Date _____

Here is 8 the 5-group way. Put an X on 2 squares. How many are left?
Fill in the number sentence and number bond.

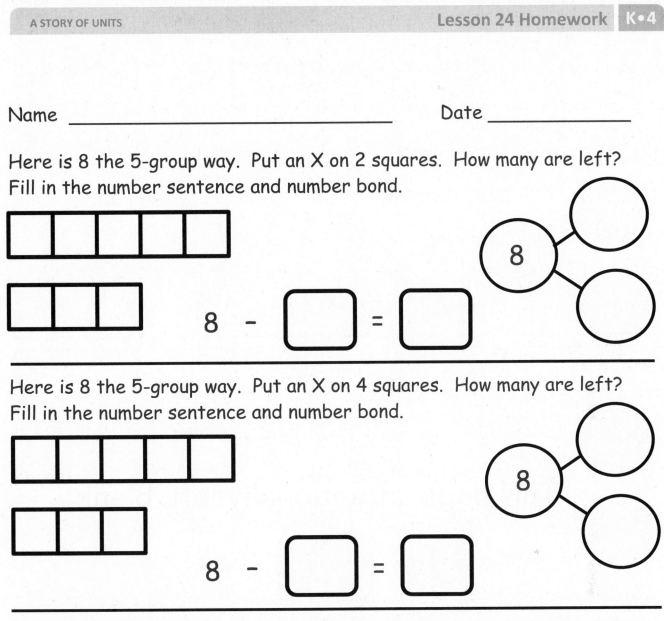

8 − ☐ = ☐

Here is 8 the 5-group way. Put an X on 4 squares. How many are left?
Fill in the number sentence and number bond.

8 − ☐ = ☐

Draw 8 the 5-group way. Put an X on some squares. How many are left?
Write the number sentence and the number bond.

8 − ☐ = ☐

On the back of your paper, draw 7 the 5-group way. Put an X on some, and
write a number sentence and number bond.

Lesson 24: Decompose the number 8 using 5-group drawings and crossing off a part, and record each decomposition with a drawing and subtraction equation.

©2015 Great Minds. eureka-math.org
GK-M4-SE-B3-1.3.1-01.2016

97

This page intentionally left blank

Name _____   Date _____

There are 9 shirts. Color some with polka dots and the rest with stripes.
Fill in the number bond.

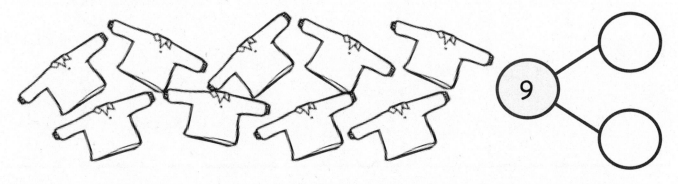

There are 9 flowers. Color some yellow and the rest red. Fill in the
number bond.

There are 9 hats. Color some brown and the rest green. Fill in the number
bond.

Lesson 25:   Model decompositions of 9 using a story situation, objects, and
            number bonds.

©2015 Great Minds. eureka-math.org
GK-M4-SE-B3-1.3.1-01.2016

99

There are 9 jellyfish. Color some blue and the rest a different color.
Fill in the number bond.

There are 9 butterflies. Color some butterflies orange and the rest a
different color. Fill in the number bond.

Draw 9 balloons. Color some red and the rest blue. Make a number bond
to match your drawing.

Lesson 25:   Model decompositions of 9 using a story situation, objects, and
number bonds.

©2015 Great Minds. eureka-math.org
GK-M4-SE-B3-1.3.1-01.2016

EUREKA
MATH™

Name _____   Date _____

There are 9 leaves. Color some of them red and the rest of them yellow.
Fill in the number bond to match.

There are 9 acorns. Color some of them green and the rest yellow. Fill in
the number bond to match.

Draw 9 birds. Color some of them blue and the rest red. Fill in the
number bond to match.

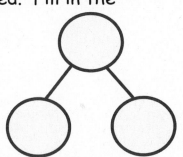

On the back of your paper, draw 9 triangles. Color some red and some
brown. Draw and fill in a number bond to match.

Lesson 25:   Model decompositions of 9 using a story situation, objects, and
number bonds.

101

©2015 Great Minds. eureka-math.org
GK-M4-SE-B3-1.3.1-01.2016

This page intentionally left blank

Name _____   Date _____

The squares below represent cube sticks.

Draw a line from the cube stick to the matching number bond. Fill in the number bond if it isn't complete.

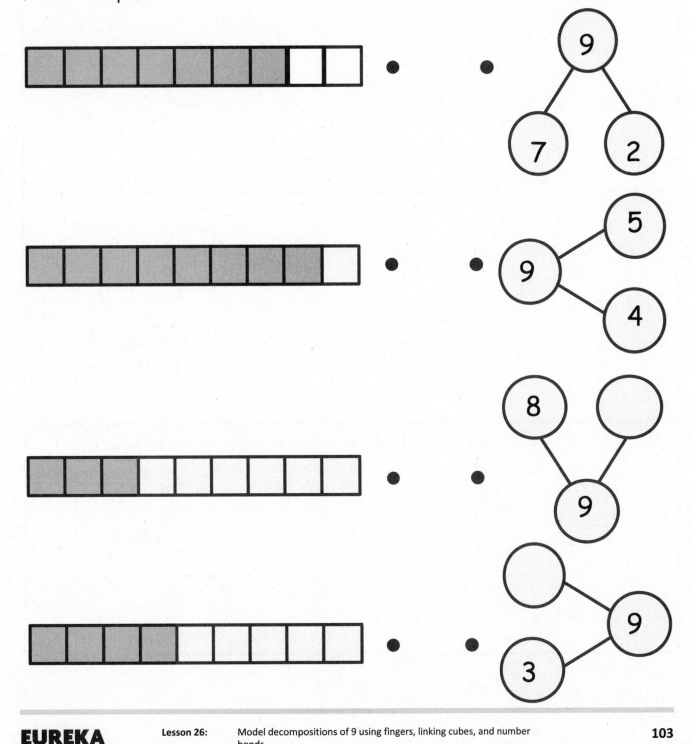

EUREKA MATH

Lesson 26:   Model decompositions of 9 using fingers, linking cubes, and number bonds.

©2015 Great Minds. eureka-math.org
GK-M4-SE-B3-1.3.1-01.2016

103

Draw and color cube sticks to match the number bonds. Fill in the number bond if it isn't complete.

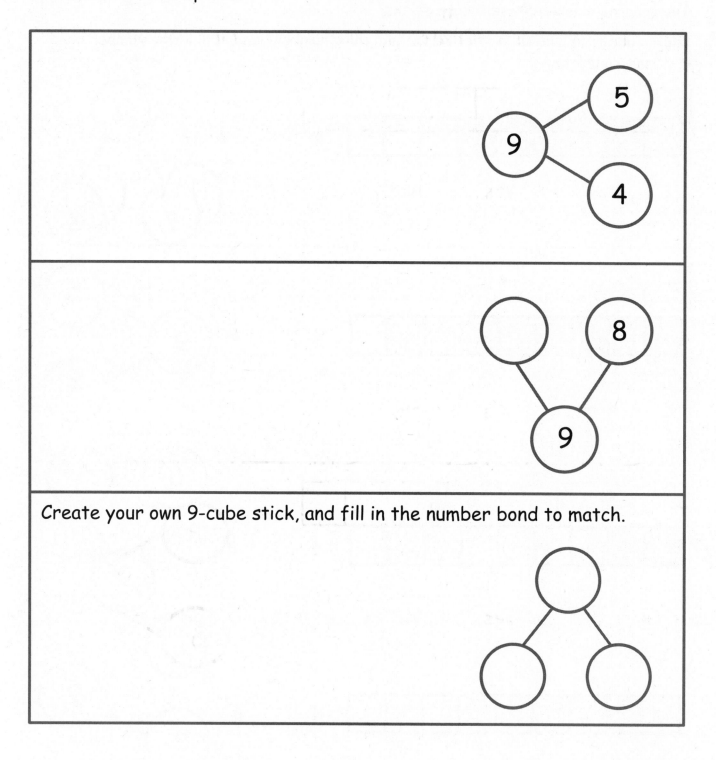

Create your own 9-cube stick, and fill in the number bond to match.

Lesson 26:   Model decompositions of 9 using fingers, linking cubes, and number bonds.

©2015 Great Minds. eureka-math.org
GK-M4-SE-B3-1.3.1-01.2016

Name _____   Date _____

The squares below represent cube sticks.
Do the linking cube sticks match the number bond? Circle yes or no.

Yes   No

Yes   No

Yes   No

Lesson 26:   Model decompositions of 9 using fingers, linking cubes, and number bonds.

©2015 Great Minds. eureka-math.org
GK-M4-SE-B3-1.3.1-01.2016

105

EUREKA MATH

Make the number bond match the cube stick.

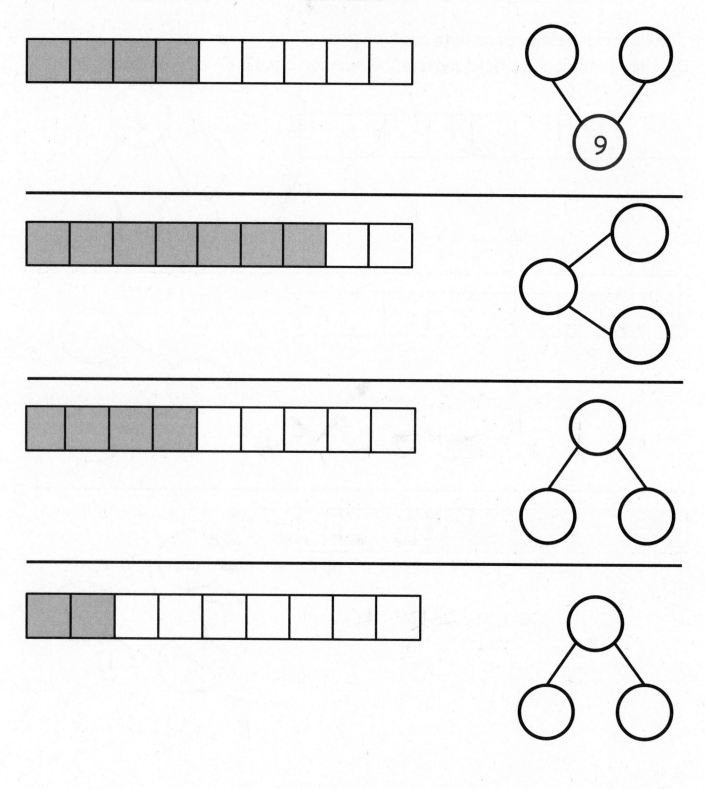

Lesson 26:   Model decompositions of 9 using fingers, linking cubes, and number bonds.

©2015 Great Minds. eureka-math.org
GK-M4-SE-B3-1.3.1-01.2016

A STORY OF UNITS

Lesson 27 Problem Set   K•4

Name _____ Date _____

Benjamin had 10 bananas. He dropped some of the bananas. Fill in the number bond to show Benjamin's bananas.

Savannah has 10 pairs of glasses. 5 are green, and the rest are purple. Color and fill in the number bond.

Xavier had 10 baseballs. Some were white, and the rest were gray. Draw the balls, and color to show how many may be white and gray. Fill in the number bond.

Lesson 27:   Model decompositions of 10 using a story situation, objects, and number bonds.

107

©2015 Great Minds. eureka-math.org
GK-M4-SE-B3-1.3.1-01.2016

There were 10 dragons playing. Some were flying, and some were running. Draw the dragons. Fill in the number bond.

Create your own story of 10. Draw your story and a number bond to go with it.

Lesson 27: Model decompositions of 10 using a story situation, objects, and number bonds.

Name _____   Date _____

Pretend this is your bracelet.
Color 5 beads blue and the rest green. Make a number bond to match.

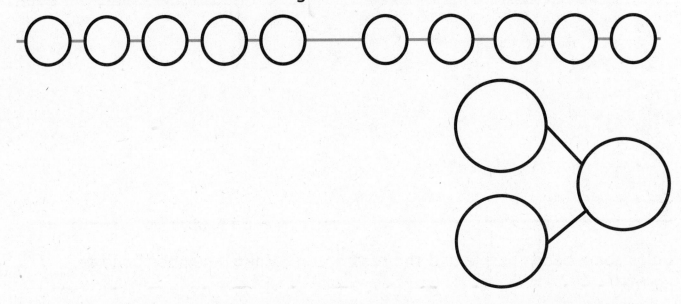

_____

Color some beads yellow and the rest orange. Make a number bond to match.

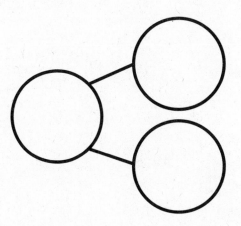

Lesson 27:   Model decompositions of 10 using a story situation, objects, and
             number bonds.

109

©2015 Great Minds. eureka-math.org
GK-M4-SE-B3-1.3.1-01.2016

Color some beads yellow and the rest black. Make a number bond to match.

◯─◯─◯─◯─◯─◯─◯─◯─◯─◯

Color some beads purple and the rest green. Make a number bond to match.

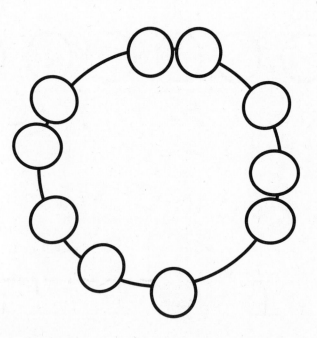

Lesson 27:   Model decompositions of 10 using a story situation, objects, and number bonds.

©2015 Great Minds. eureka-math.org
GK-M4-SE-B3-1.3.1-01.2016

Name _____   Date _____

These squares represent cube sticks. Look at the linking cube sticks.
Draw a line from the cube stick to the number bond that matches.
Fill in the number bond if it is not complete.

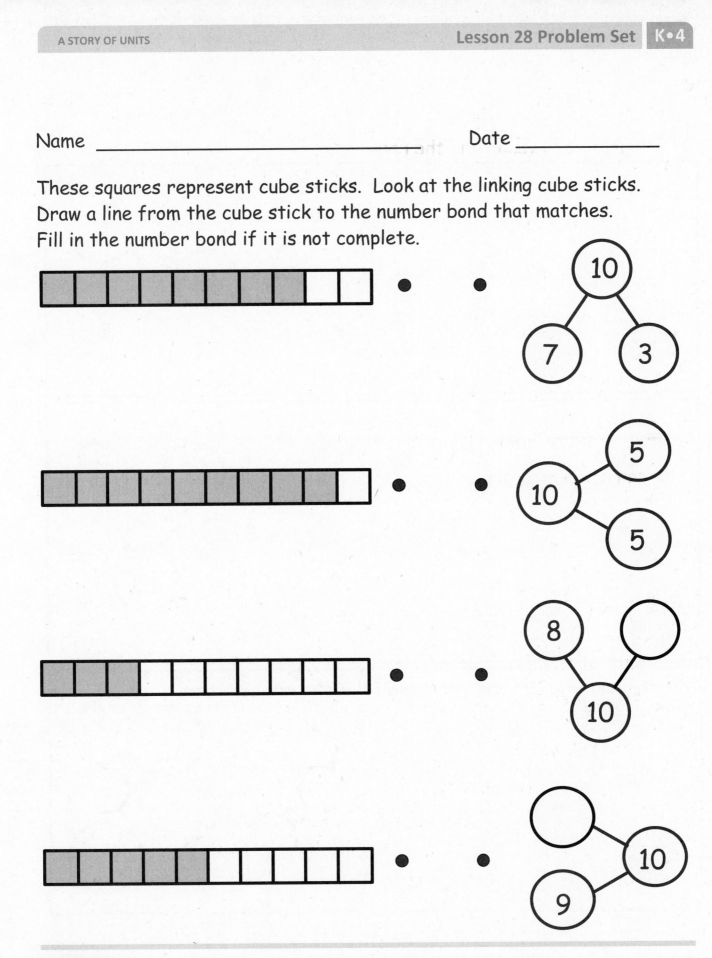

EUREKA MATH

Lesson 28:   Model decompositions of 10 using fingers, sets, linking cubes, and
number bonds.

©2015 Great Minds. eureka-math.org
GK-M4-SE-B3-1.3.1-01.2016

111

Draw and color cube sticks to match the number bonds.

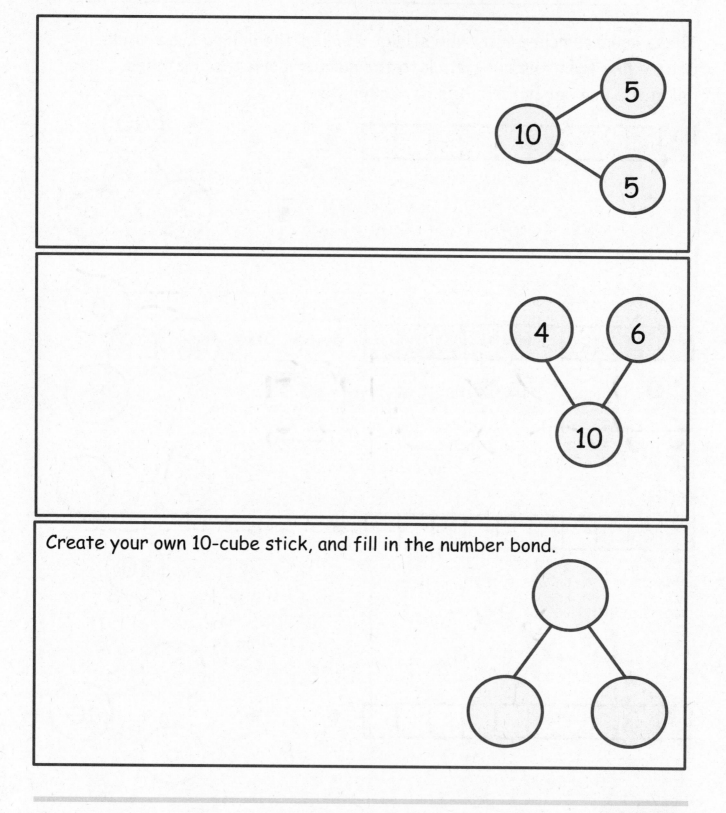

Create your own 10-cube stick, and fill in the number bond.

Lesson 28:   Model decompositions of 10 using fingers, sets, linking cubes, and number bonds.

©2015 Great Minds. eureka-math.org
GK-M4-SE-B3-1.3.1-01.2016

EUREKA MATH™

Name _____   Date _____

Write a number bond to match each domino.

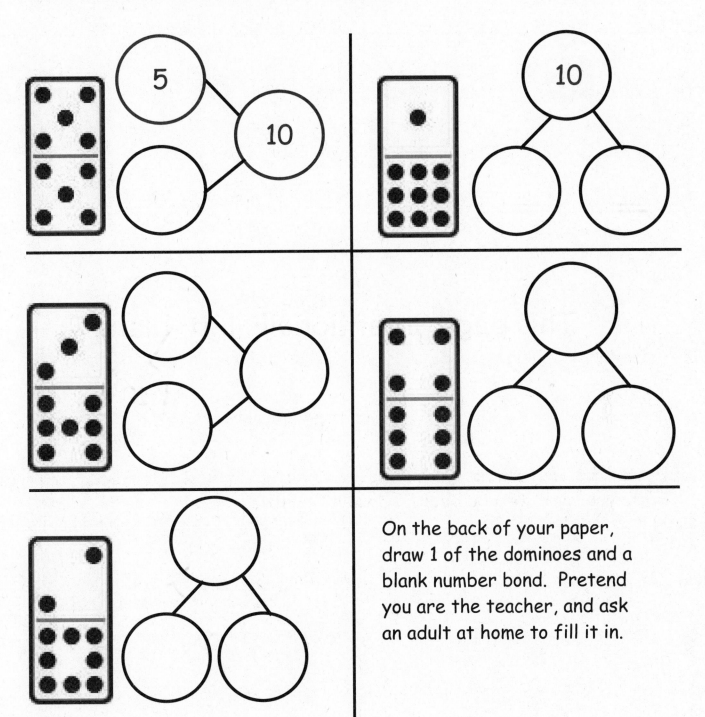

On the back of your paper, draw 1 of the dominoes and a blank number bond. Pretend you are the teacher, and ask an adult at home to fill it in.

EUREKA MATH

Lesson 28:   Model decompositions of 10 using fingers, sets, linking cubes, and number bonds.

113

©2015 Great Minds. eureka-math.org
GK-M4-SE-B3-1.3.1-01.2016

This page intentionally left blank

Name _____   Date _____

Izzy had a tea party with 7 teddy bears and 2 dolls. There were 9 friends at the party. Fill in the number bond and number sentence.

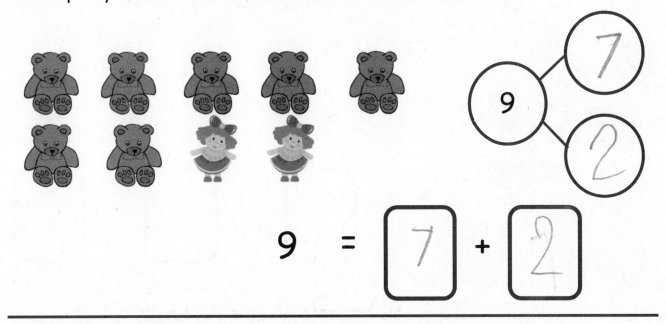

$$9 \quad = \quad \boxed{7} \quad + \quad \boxed{2}$$

Robin had 9 vegetables on her plate. She had 3 carrots and 6 peas. Draw the carrots and peas in the 5-group way. Fill in the number sentence.

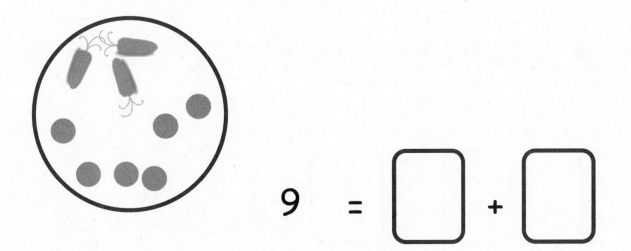

$$9 \quad = \quad \boxed{} \quad + \quad \boxed{}$$

Lesson 29:   Represent pictorial decomposition and composition addition stories
to 9 with 5-group drawings and equations with no unknown.

115

©2015 Great Minds. eureka-math.org
GK-M4-SE-B3-1.3.1-01.2016

Shane played with 5 toy zebras and 4 toy lions. He had 9 animal toys in all.
Draw black and tan circles to show the zebras and the lions in the 5-group
way. Fill in the number sentence.

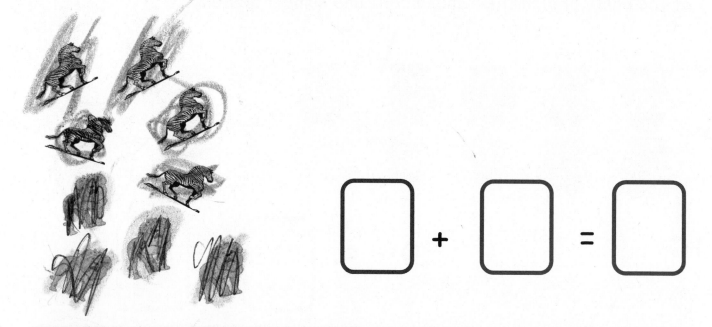

Jimmy had 9 marbles. 8 were red, and 1 was green. Draw the marbles in
the 5-group way. Fill in the number bond and number sentence.

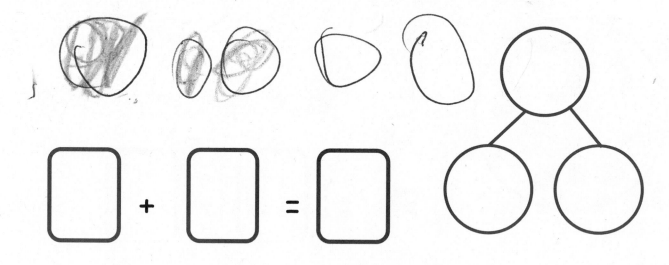

Lesson 29: Represent pictorial decomposition and composition addition stories
to 9 with 5-group drawings and equations with no unknown.

©2015 Great Minds. eureka-math.org
GK-M4-SE-B3-1.3.1-01.2016

EUREKA MATH™

Name _____   Date _____

Jack found 7 balls while cleaning the toy bin. He found 6 basketballs and 1 baseball. Fill in the number sentence and the number bond.

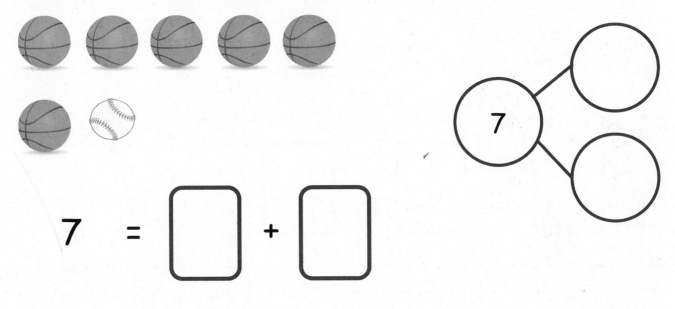

7 = [   ] + [   ]

_____

Jack found 7 mitts and 2 bats. He found 9 things. Fill in the number sentence and the number bond.

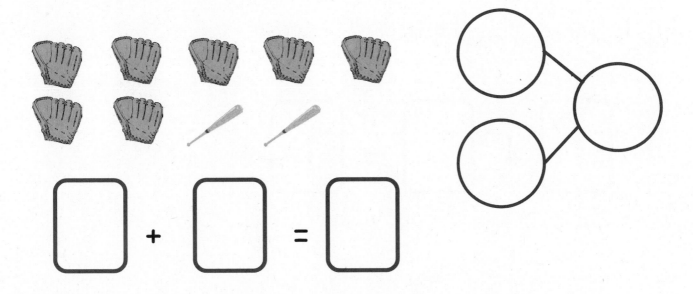

[   ] + [   ] = [   ]

Lesson 29:   Represent pictorial decomposition and composition addition stories
              to 9 with 5-group drawings and equations with no unknown.

117

©2015 Great Minds. eureka-math.org
GK-M4-SE-B3-1.3.1-01.2016

Jack found 8 hockey pucks and 1 hockey stick. He found 9 hockey things.
Draw the hockey pucks and stick in the 5-group way. Fill in the number
sentence.

Jack needs a snack. He found 9 pieces of fruit. 5 were strawberries, and
4 were grapes. Draw the strawberries and grapes in the 5-group way.
Fill in the number sentence.

Lesson 29:   Represent pictorial decomposition and composition addition stories
                      to 9 with 5-group drawings and equations with no unknown.

©2015 Great Minds. eureka-math.org
GK-M4-SE-B3-1.3.1-01.2016

Name _____   Date _____

Fill in the number bonds, and complete the number sentences.

Ricky has 10 space toys. He has 7 rockets and 3 astronauts.

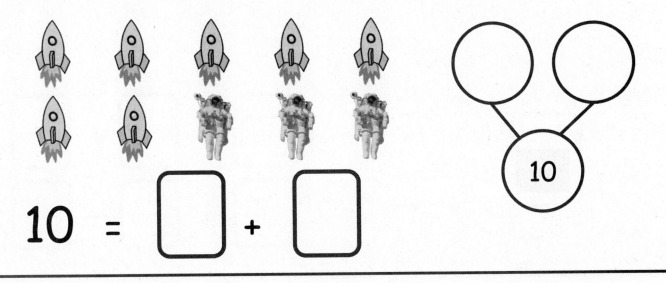

10 = [   ] + [   ]

Bianca has 4 pigs and 6 sheep on her farm. She has 10 animals altogether.

[   ] + [   ] = [   ]

Lesson 30:   Represent pictorial decomposition and composition addition
stories to 10 with 5-group drawings and equations with no unknown.

119

©2015 Great Minds. eureka-math.org
GK-M4-SE-B3-1.3.1-01.2016

Danica had 5 green balloons. Her friend gave her 5 blue balloons. Draw all of her balloons in the 5-group way. Fill in both number sentences.

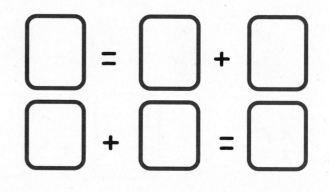

Jason is playing with 10 bouncy balls. He has 8 on the table and 2 on the floor. Draw the bouncy balls in the 5-group way. Fill in both number sentences.

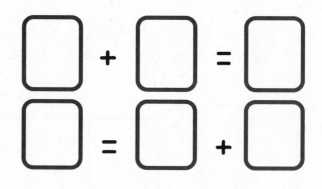

Lesson 30:   Represent pictorial decomposition and composition addition
               stories to 10 with 5-group drawings and equations with no unknown.

©2015 Great Minds. eureka-math.org
GK-M4-SE-B3-1.3.1-01.2016

Name _____   Date _____

Fill in the number bonds, and complete the number sentences.

Scott went to the zoo. He saw 6 giraffes and 4 zebras. He saw 10 animals altogether.

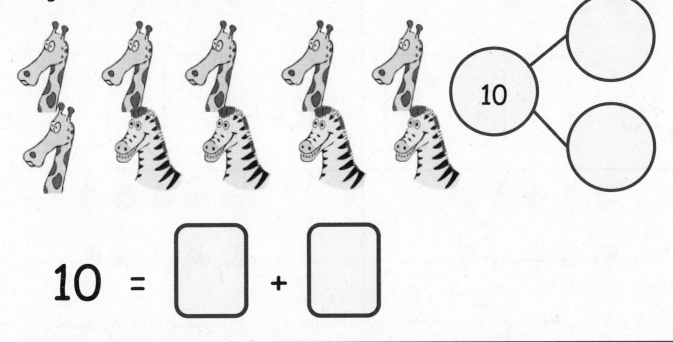

10 = [   ] + [   ]

Susan saw 10 animals at the zoo. She saw 5 lions and 5 elephants.
Draw the animals in the 5-group way.

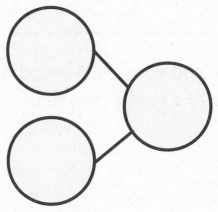

[   ] + [   ] = [   ]

Lesson 30:   Represent pictorial decomposition and composition addition
stories to 10 with 5-group drawings and equations with no unknown.

©2015 Great Minds. eureka-math.org
GK-M4-SE-B3-1.3.1-01.2016

121

Make 2 groups. Circle 1 of the groups. Write a number sentence to match. Find as many partners of 10 as you can.

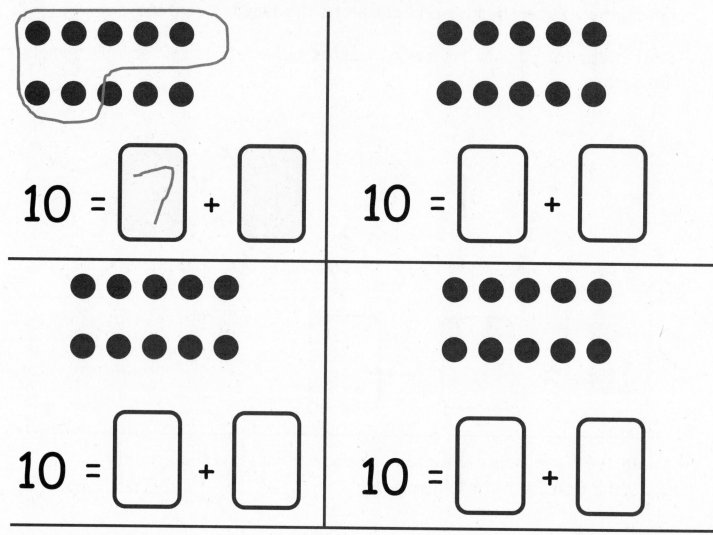

$$10 = \boxed{7} + \boxed{}$$

$$10 = \boxed{} + \boxed{}$$

$$10 = \boxed{} + \boxed{}$$

$$10 = \boxed{} + \boxed{}$$

Draw 10 dots the 5-group way. Make 2 groups. Circle one of the groups. Write a number sentence to match your drawing.

Lesson 30:   Represent pictorial decomposition and composition addition
              stories to 10 with 5-group drawings and equations with no unknown.

EUREKA MATH™

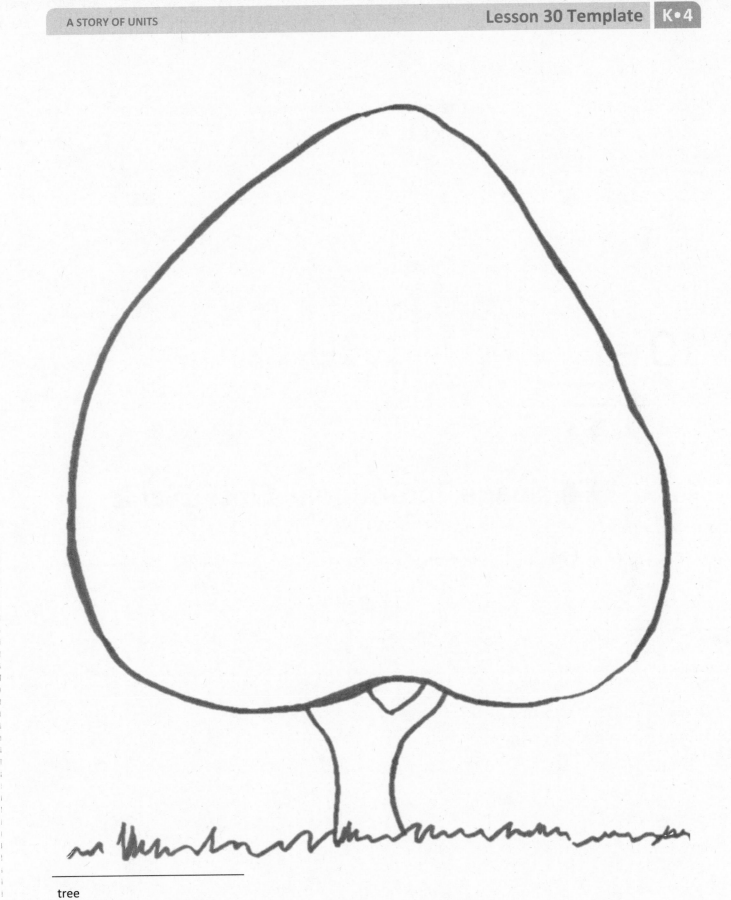

tree

Lesson 30:   Represent pictorial decomposition and composition addition
stories to 10 with 5-group drawings and equations with no unknown.

123

©2015 Great Minds. eureka-math.org
GK-M4-SE-B3-1.3.1-01.2016

This page intentionally left blank

Name _____   Date _____

Draw the story. Fill in the number sentence.

Zayne had 6 round crackers and 3 square crackers. How many crackers did Zayne have in all?

_____   +   _____   =   _____

Riley had 9 crayons. Her friend gave her 1 crayon. How many crayons did Riley have in all?

_____   +   _____   =   _____

Lesson 31:   Solve *add to with total unknown* and *put together with total unknown* problems with totals of 9 and 10.

©2015 Great Minds. eureka-math.org
GK-M4-SE-B3-1.3.1-01.2016

125

Draw the story.   Write a number sentence to match.

Jenny had 3 red and 7 purple pieces of construction paper. How many pieces of construction paper did Jenny have altogether?

Rhett had 5 square blocks. His friend gave him 4 rectangle blocks. How many blocks did Rhett have altogether?

Lesson 31:   *Solve add to with total unknown* and *put together with total unknown* problems with totals of 9 and 10.

Name _____   Date _____

Draw the story.   Fill in the number sentence.

Jake has 7 chocolate cookies and 2 sugar cookies. How many cookies does he have altogether?

_____ + _____ = _____

Jake's mother bought juice boxes. 4 were apple juice, and 5 were orange juice. How many juice boxes did she have in all?

_____ + _____ = _____

Lesson 31:   Solve *add to with total unknown* and *put together with total unknown* problems with totals of 9 and 10.

©2015 Great Minds. eureka-math.org
GK-M4-SE-B3-1.3.1-01.2016

127

Draw the story. Write a number sentence to match.

Ryan had 5 celery sticks and 5 carrot sticks. How many veggie sticks did Ryan have altogether?

Draw an addition story, and write a number sentence to match it. Explain your work to an adult at home.

Lesson 31:   Solve *add to with total unknown* and *put together with total unknown* problems with totals of 9 and 10.

©2015 Great Minds. eureka-math.org
GK-M4-SE-B3-1.3.1-01.2016

EUREKA
MATH

|
=
|

+

|

_____

equation

Lesson 31:   Solve *add to with total unknown* and *put together with total unknown* problems with totals of 9 and 10.

129

©2015 Great Minds. eureka-math.org
GK-M4-SE-B3-1.3.1-01.2016

This page intentionally left blank

Name _____ Date _____

Listen to the word problem. Fill in the number sentence.

Cecilia has 9 bows. Some have polka dots, and some have stripes. How many polka dot and how many striped bows do you think Cecilia has?

$$9 = \boxed{} + \boxed{}$$

Keegan has 10 train cars. Some are black, and some are green. How many black and green train cars do you think Keegan has?

$$10 = \boxed{} + \boxed{}$$

Lesson 32: Solve *both addends unknown* word problems with totals of 9 and 10 using 5-group drawings.

131

©2015 Great Minds. eureka-math.org
GK-M4-SE-B3-1.3.1-01.2016

Kate has 9 heart stickers. Some are yellow, and the rest are green. Show two different ways Kate's stickers could look. Fill in the number sentences to match.

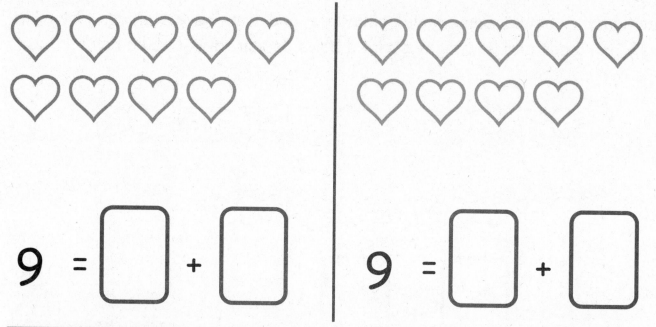

$$9 = \boxed{} + \boxed{}$$

$$9 = \boxed{} + \boxed{}$$

Danny has 10 robots. Some are red, and the rest are gray. Show two different ways Danny's robots could look. Fill in the number sentences to match.

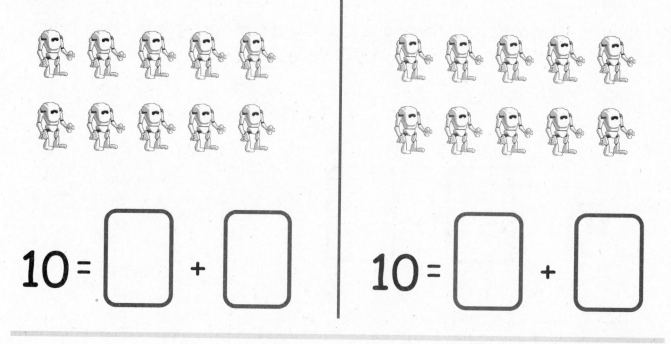

$$10 = \boxed{} + \boxed{}$$

$$10 = \boxed{} + \boxed{}$$

Lesson 32: Solve *both addends unknown* word problems with totals of 9 and 10 using 5-group drawings.

Name _____   Date _____

Color the robots to match the number sentence. Tell a story about the robots.

10 = 5 + 5

_____

10 = 6 + 4

_____

10 = 7 + 3

_____

10 = 8 + 2

_____

10 = 9 + 1

©2015 Great Minds. eureka-math.org
GK-M4-SE-B3-1.3.1-01.2016

Name _____   Date _____

Jerry has 9 baseball hats. Draw the hats the 5-group way. Color some red and some blue. Fill in the number sentence to match.

$$9 = \boxed{} + \boxed{}$$

_____

Anne had 10 pencils. Draw the pencils the 5-group way. Color some pencils blue and some yellow. Fill in the number sentence to match.

$$10 = \boxed{} + \boxed{}$$

Lesson 32:   Solve *both addends unknown* word problems with totals of 9 and 10 using 5-group drawings.

There are 10 apples. Color some red and the rest green. Then, show a different way the apples could look. Fill in the number sentences to match.

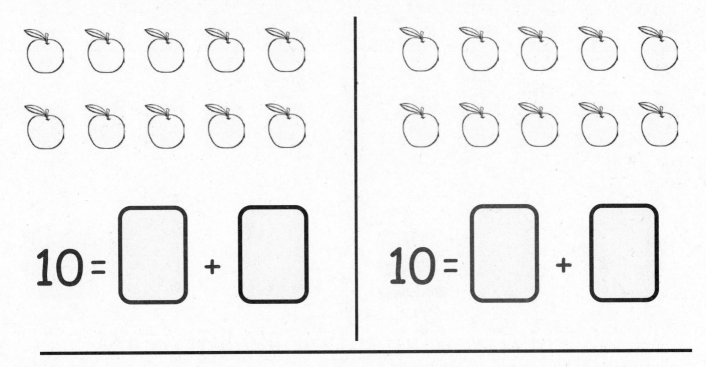

10 = ☐ + ☐   10 = ☐ + ☐

Anya has 9 stuffed cats. Some are orange, and the rest are gray. Show two different ways Anya's cats could look. Fill in the number sentences to match.

9 = ☐ + ☐   9 = ☐ + ☐

Lesson 32:   Solve *both addends unknown* word problems with totals of 9 and 10 using 5-group drawings.

135

This page intentionally left blank

Name _____   Date _____

Fill in the number sentence to match the story.

There were 7 trains. 2 trains rolled away. Now there are 5 trains.

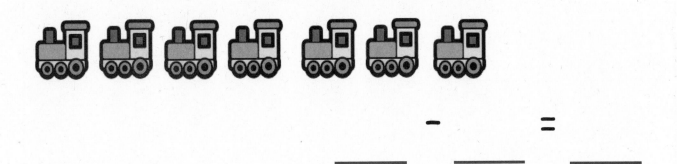

_____ - _____ = _____

There were 9 cars at the stop sign. 7 drove away. There are 2 cars left.

_____ - _____ = _____

There were 10 people. 6 people got on the bus. Now there are 4 people.

_____ - _____ = _____

Draw the story. Fill in the number sentence to match.

The bus had 10 people. 5 people got off. Now there are 5 people left.

____ - ____ = ____

There were 9 planes in the sky. 3 planes landed. Now there are 6 planes in the sky.

____ - ____ = ____

Lesson 33:   Solve *take from* equations with no unknown using numbers to 10.

Name _____   Date _____

Fill in the number sentence and the number bond.

There were 10 teddy bears. Cross out 2 bears. There are 8 bears left.

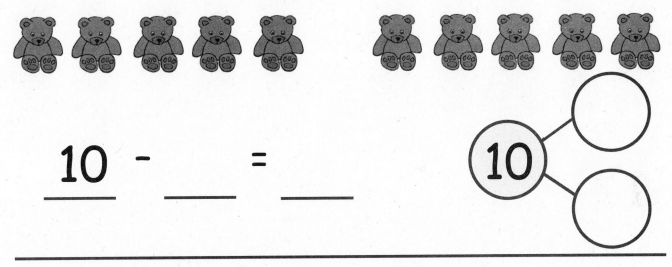

10 - ___ = ___

There were 10 teddy bears. Cross out 9. There is 1 left.

___ - ___ = ___

There were 10 teddy bears. Cross out 3. There are 7 bears left.

___ - ___ = ___

EUREKA MATH

Draw a line from the picture to the number sentence it matches.

• 10 – 1 = 9

• 10 – 3 = 7

• 9 – 4 = 5

• 9 – 8 = 1

Solve *take from* equations with no unknown using numbers to 10.

EUREKA
MATH™

_____

subtraction equation

$$= $$

Lesson 33:   Solve *take from* equations with no unknown using numbers to 10.

141

©2015 Great Minds. eureka-math.org
GK-M4-SE-B3-1.3.1-01.2016

This page intentionally left blank

Name _____   Date _____

Fill in the number sentences and number bonds.

There are 9 babies playing. 2 crawl away. How many babies are left?

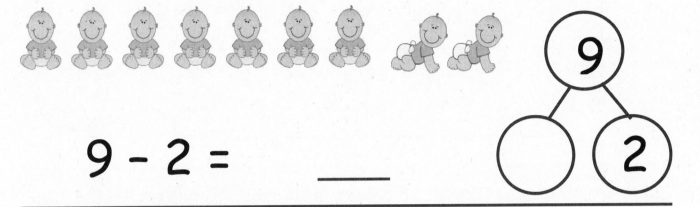

9 – 2 = _____

There are 10 babies playing. 1 crawls away. How many babies are left?

10 – _____ = _____

There are 9 babies playing. 6 crawl away. How many babies are left?

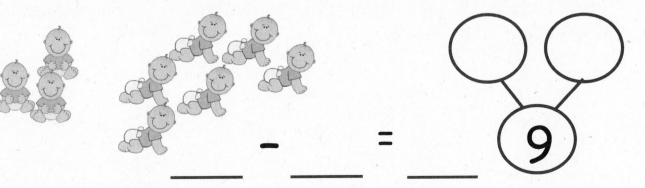

_____ – _____ = _____

Lesson 34:   Represent subtraction story problems by breaking off, crossing out, and hiding a part.

143

The squares below represent cube sticks.
Carlos had a 9-stick. He broke off 4 cubes to share with his friend.
How many cubes are left? Draw a line to show where he broke his stick.

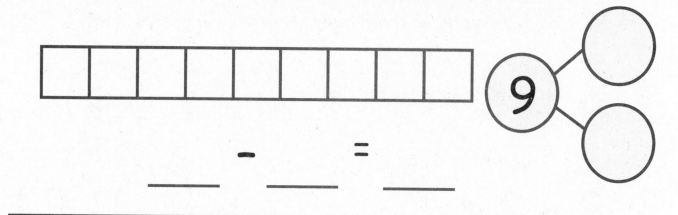

_____ - _____ = _____

Sophie had 10 grapes. She ate 6 grapes. How many grapes are left?
Draw her grapes, and cross off the ones she ate.

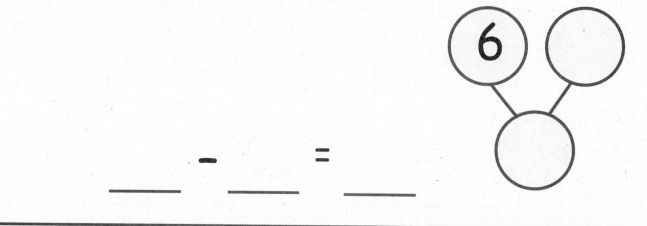

_____ - _____ = _____

Spot had 10 bones. He hid 8 bones in the ground. How many bones does he have now? Draw Spot's bones.

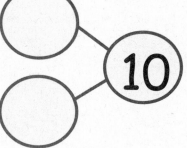

_____ - _____ = _____

Lesson 34:   Represent subtraction story problems by breaking off, crossing out, and hiding a part.

©2015 Great Minds. eureka-math.org
GK-M4-SE-B3-1.3.1-01.2016

EUREKA MATH

Name _____   Date _____

There were 8 penguins. 2 penguins went back to the ship. Cross out 2 penguins. Fill in the number sentence and the number bond.

8 – 2 = _____

The squares below represent cubes.
Count the cubes. Draw a line to break 4 cubes off the train. Fill in the number sentence and the number bond.

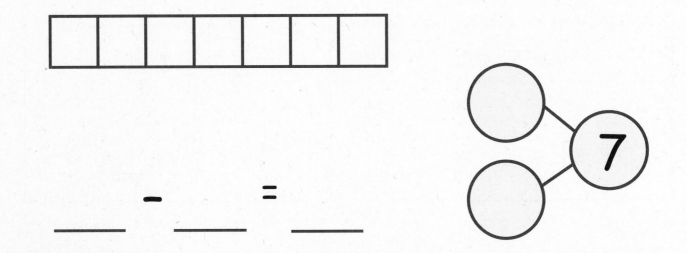

___ – ___ = ___

There are 10 bears. Some go inside the cave to hide. Cross them out.
Complete the number sentence.

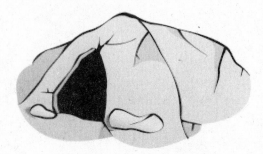

$$10 - \underline{\quad} = \underline{\quad}$$

Complete these number sentences.

$5 - 1 = \boxed{}$

$\boxed{} = 2 + 3$

$\boxed{} = 5 - 4$

$2 + 2 = \boxed{}$

Complete these number sentences.

$3 - 1 = \boxed{}$

$\boxed{} = 1 + 3$

$\boxed{} = 4 - 2$

$1 + 2 = \boxed{}$

Lesson 34: Represent subtraction story problems by breaking off, crossing out, and hiding a part.

Name _____   Date _____

Cross off the part that goes away. Fill in the number bond and number sentence.

Jeremy had 9 baseballs. He took 5 baseballs outside to play, and they got lost. How many balls are left?

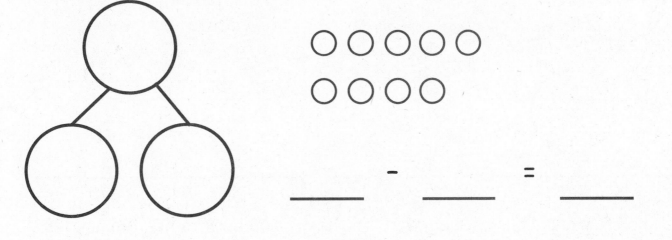

_____ - _____ = _____

Sandy had 9 leaves. Then, 4 leaves blew away. How many leaves are left?

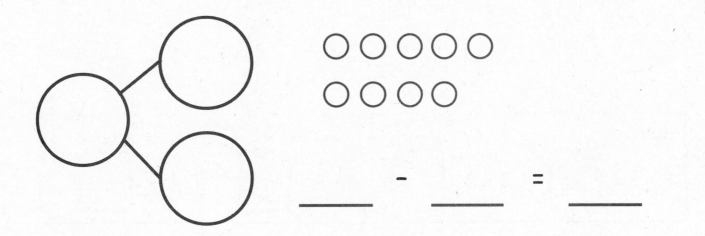

_____ - _____ = _____

EUREKA MATH

Lesson 35:   Decompose the number 9 using 5-group drawings, and record each decomposition with a subtraction equation.

©2015 Great Minds. eureka-math.org
GK-M4-SE-B3-1.3.1-01.2016

147

Make a 5-group drawing to show the story. Cross off the part that goes away. Fill in the number bond and number sentence.

Ryder had 9 star stickers. He gave 3 to his friend. How many star stickers does Ryder have now?

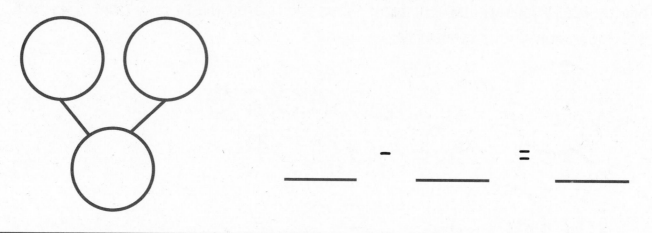

_____ - _____ = _____

Jen had 9 granola bars. She gave 8 of the granola bars to her teammates. How many granola bars does she have left?

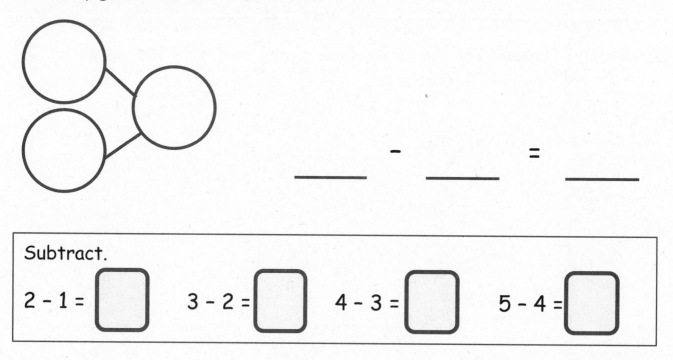

_____ - _____ = _____

Subtract.

2 – 1 = ☐   3 – 2 = ☐   4 – 3 = ☐   5 – 4 = ☐

Lesson 35:   Decompose the number 9 using 5-group drawings, and record each decomposition with a subtraction equation.

EUREKA MATH™

Name _____   Date _____

Cross off the part that goes away. Fill in the number bond and number sentence.

Mary had 9 library books. She returned 1 book to the library. How many books are left?

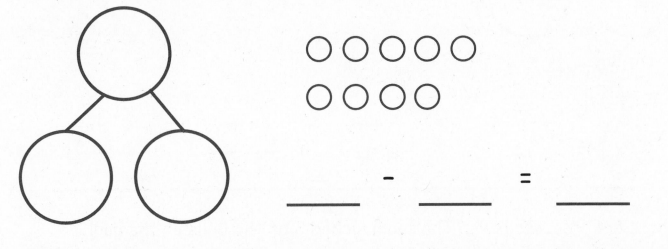

_____ - _____ = _____

There were 9 lunch bags. 3 bags were thrown away. How many bags are there now?

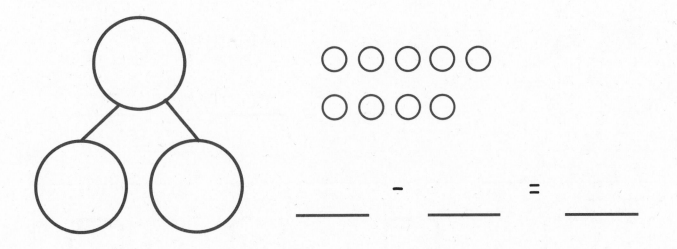

_____ - _____ = _____

Lesson 35:   Decompose the number 9 using 5-group drawings, and record each
             decomposition with a subtraction equation.

©2015 Great Minds. eureka-math.org
GK-M4-SE-B3-1.3.1-01.2016   149

Make a 5-group drawing to show the story. Cross off the part that goes away. Fill in the number bond and number sentence.

Ms. Lopez has 9 pencils. 7 of them broke. How many pencils are left?

_____ - _____ = _____

There are 9 soccer balls. The team kicked 5 of the balls at the goal. How many soccer balls are left?

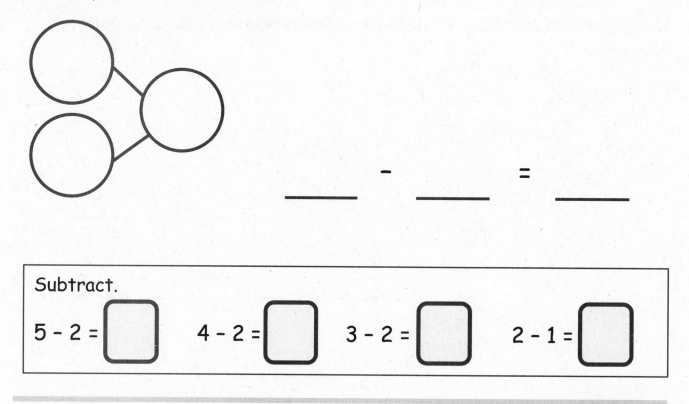

_____ - _____ = _____

Subtract.

5 – 2 = ☐   4 – 2 = ☐   3 – 2 = ☐   2 – 1 = ☐

Lesson 35:   Decompose the number 9 using 5-group drawings, and record each decomposition with a subtraction equation.

©2015 Great Minds. eureka-math.org
GK-M4-SE-B3-1.3.1-01.2016

Name _____   Date _____

Fill in the number bond and number sentence.   Cross off the part that goes away.

Stan had 10 blueberries. He ate 5 berries. How many blueberries are left?

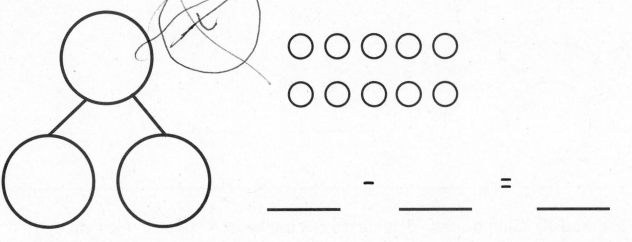

_____   _____   =   _____

Tracy had 10 heart stickers. She lost 1 sticker. How many stickers are left?

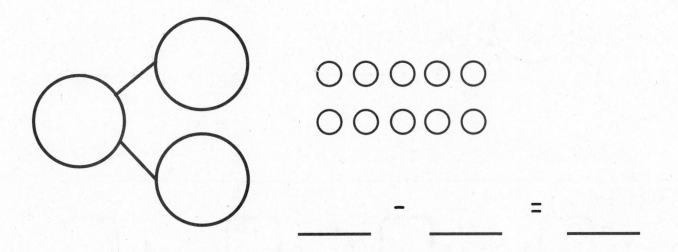

_____   _____   =   _____

EUREKA MATH

Lesson 36:   Decompose the number 10 using 5-group drawings, and record each decomposition with a subtraction equation.

151

©2015 Great Minds. eureka-math.org
GK-M4-SE-B3-1.3.1-01.2016

Make a 5-group drawing to show the story. Fill in the number bond and number sentence. Cross off the part that goes away.

Nick had 10 party hats. 7 hats were thrown away. How many hats does Nick have now?

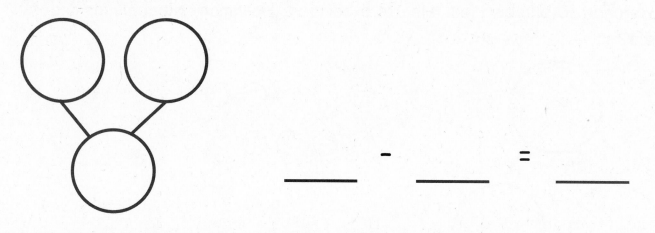

_____ − _____ = _____

Tatiana had 10 juice boxes. 3 juice boxes broke and spilled. How many full juice boxes does she have left?

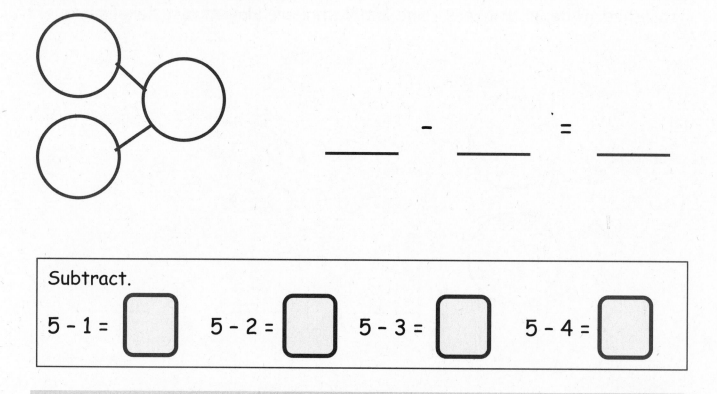

_____ − _____ = _____

Subtract.

5 − 1 = ☐   5 − 2 = ☐   5 − 3 = ☐   5 − 4 = ☐

Lesson 36:   Decompose the number 10 using 5-group drawings, and record each decomposition with a subtraction equation.

©2015 Great Minds. eureka-math.org
GK-M4-SE-B3-1.3.1-01.2016

EUREKA MATH

Name _Lwt shuhat_   Date _____

Fill in the number bond and number sentence.   Cross off the part that goes away.

MacKenzie had 10 buttons on her jacket. 2 buttons broke off her jacket. How many buttons are left on her jacket?

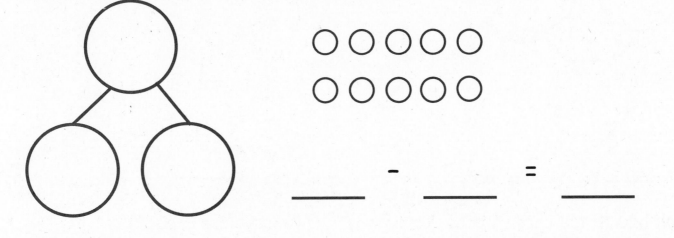

_____ - _____ = _____

Donna had 10 cups. 6 cups fell and broke. How many unbroken cups are there now?

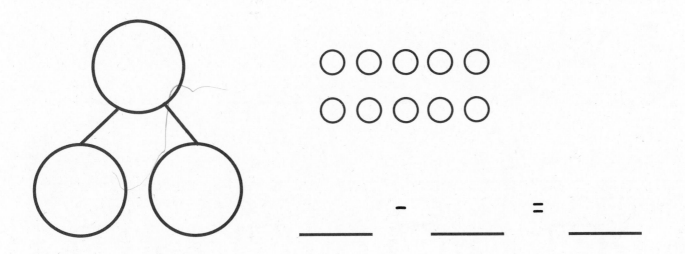

_____ - _____ = _____

Lesson 36:   Decompose the number 10 using 5-group drawings, and record each decomposition with a subtraction equation.

153

©2015 Great Minds. eureka-math.org
GK-M4-SE-B3-1.3.1-01.2016

Make a 5-group drawing to show the story. Fill in the number bond and number sentence. Cross off the part that goes away.

There were 10 butterflies. 9 butterflies flew away. How many are left?

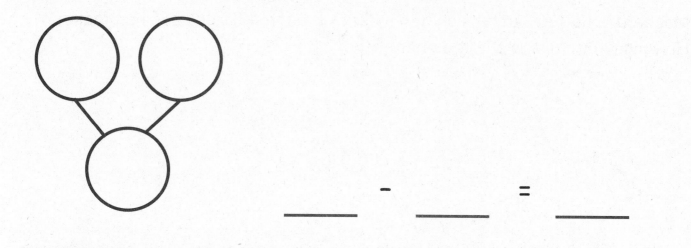

_____ - _____ = _____

Bob had 10 toy cars. 4 cars drove away. How many cars are left?

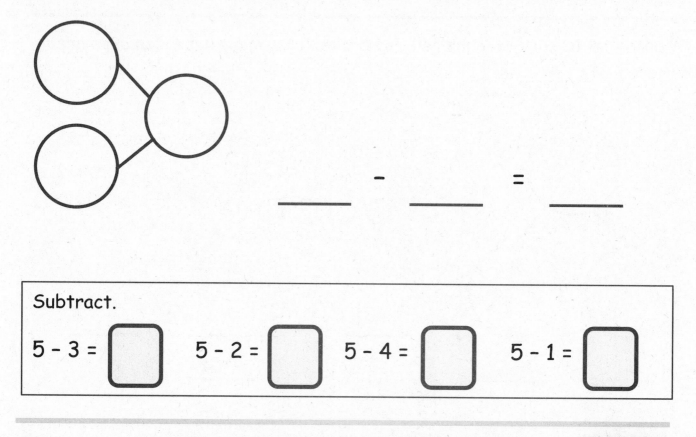

_____ - _____ = _____

Subtract.

5 – 3 = ☐   5 – 2 = ☐   5 – 4 = ☐   5 – 1 = ☐

Lesson 36:   Decompose the number 10 using 5-group drawings, and record each decomposition with a subtraction equation.

EUREKA MATH

Name _____ Date _____

Listen to each story. Show the story with your fingers on the number path. Then, fill in the number sentence.

1	2	3	4	5	6	7	8	9	10

Freddy had 3 strawberries for a snack. His dad gave him 2 more strawberries. How many strawberries does Freddy have now?

$$\underline{} 3 + \underline{} 2 = \underline{}$$

Freddy ate 2 of his strawberries. How many strawberries does Freddy have now?

$$\underline{} 5 - \underline{} 2 = \underline{}$$

Logan had 7 frogs. 2 frogs hopped away. How many frogs does Logan have now?

$$\underline{} - \underline{} = \underline{}$$

Pretend that Logan's 2 frogs hopped back. How many frogs does he have now?

$$\underline{} + \underline{} = \underline{}$$

©2015 Great Minds. eureka-math.org
GK-M4-SE-B3-1.3.1-01.2016

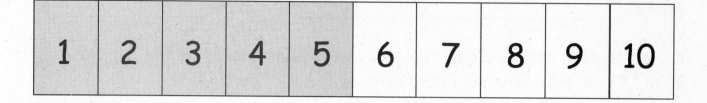

Stella had 4 pennies. She found 3 more pennies. How many pennies does Stella have now?

_____ + _____ = _____

Stella gave the 3 pennies to her dad. How many pennies does she have now?

_____ - _____ = _____

Marissa made 8 bracelets. She loved them so much she did not give any away. How many bracelets does Marissa have now?

_____ - _____ = _____

Jackson found 6 toys under his bed. He looked and did not find any more toys. How many toys does Jackson have now?

_____ + _____ = _____

Solve.

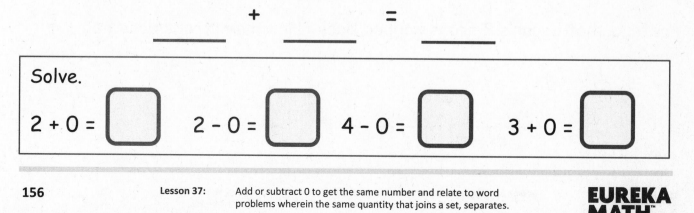

$2 + 0 =$ ☐ $2 - 0 =$ ☐ $4 - 0 =$ ☐ $3 + 0 =$ ☐

Lesson 37:   Add or subtract 0 to get the same number and relate to word problems wherein the same quantity that joins a set, separates.

©2015 Great Minds. eureka-math.org
GK-M4-SE-B3-1.3.1-01.2016

EUREKA MATH™

Name _____   Date _____

Listen to each story. Show the story with your fingers on the number path. Then, fill in the number sentence and number bond.

1	2	3	4	5	6	7	8	9	10

Joey had 5 pennies. He found 3 pennies in the couch. How many pennies does Joey have now?

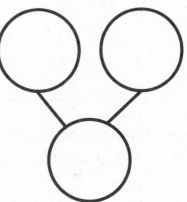

_____ + _____ = _____

Joey gave the 3 pennies to his dad. How many pennies does Joey have now?

_____ – _____ = _____

Lesson 37:   Add or subtract 0 to get the same number and relate to word problems wherein the same quantity that joins a set, separates.

©2015 Great Minds. eureka-math.org
GK-M4-SE-B3-1.3.1-01.2016

157

| 1 | 2 | 3 | 4 | 5 | 6 | 7 | 8 | 9 | 10 |

Siri had 9 pennies. She looked all around the house but could not find any more pennies. How many pennies does she have now?

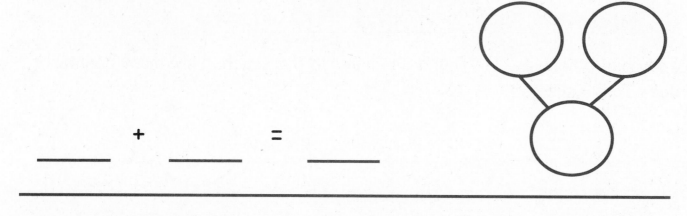

_____ + _____ = _____

There were 8 children waiting for the school bus. No more children came to the bus stop. How many children are waiting now?

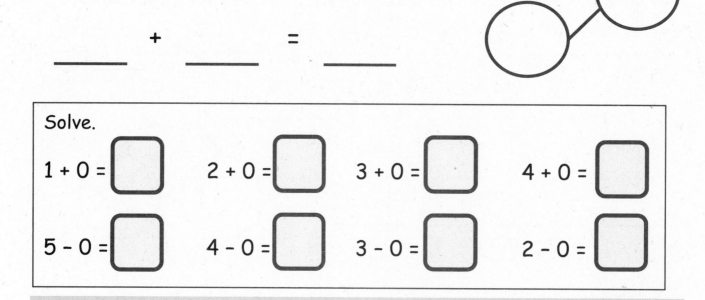

_____ + _____ = _____

Solve.

1 + 0 = ☐   2 + 0 = ☐   3 + 0 = ☐   4 + 0 = ☐

5 − 0 = ☐   4 − 0 = ☐   3 − 0 = ☐   2 − 0 = ☐

Lesson 37:   Add or subtract 0 to get the same number and relate to word problems wherein the same quantity that joins a set, separates.

Name _____   Date _____

| 1 | 2 | 3 | 4 | 5 | 6 | 7 | 8 | 9 | 10 |

Use the number path to add. Write the number in the box. Color the circles to match. Use a different color to show 1 more.

1 + 1 = [ ]

2 + 1 = [ ]

3 + 1 = [ ]

4 + 1 = [ ]

5 + 1 = [ ]

EUREKA MATH

Lesson 38:   Add 1 to numbers 1–9 to see the pattern of *the next number* using 5-group drawings and equations.

159

©2015 Great Minds. eureka-math.org
GK-M4-SE-B3-1.3.1-01.2016

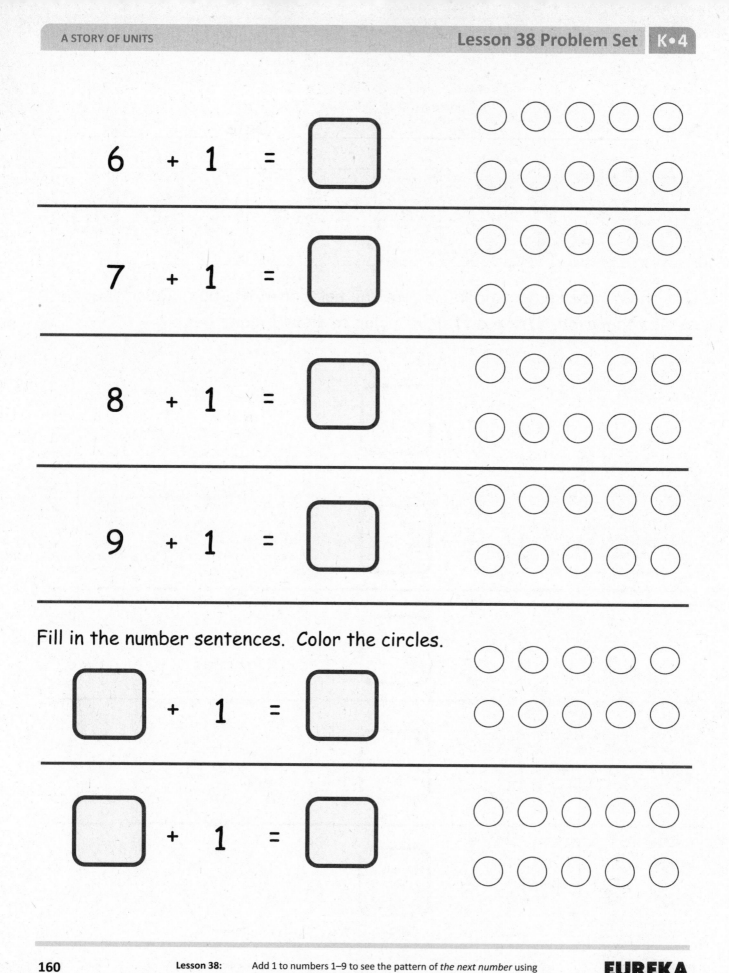

6 + 1 = ☐

7 + 1 = ☐

8 + 1 = ☐

9 + 1 = ☐

Fill in the number sentences. Color the circles.

☐ + 1 = ☐

☐ + 1 = ☐

Lesson 38: Add 1 to numbers 1–9 to see the pattern of *the next number* using
5-group drawings and equations.

©2015 Great Minds. eureka-math.org
GK-M4-SE-B3-1.3.1-01.2016

EUREKA
MATH

Name _____ Date _____

Follow the instructions to color the 5-group. Then, fill in the number sentence or number bond to match.

Color 9 squares green and 1 square blue.

_____ + _____ = _____

Color 8 squares green and 1 square blue.

_____ + _____ = _____

Color 7 squares green and 1 square blue.

_____ + _____ = _____

Lesson 38: Add 1 to numbers 1–9 to see the pattern of *the next number* using 5-group drawings and equations.

161

EUREKA MATH™

Color 2 squares green and 1 square blue.

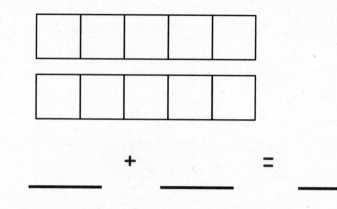

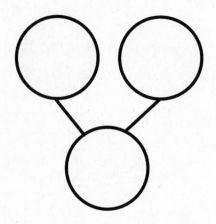

_____ + _____ = _____

Color 1 square green and 1 square blue.

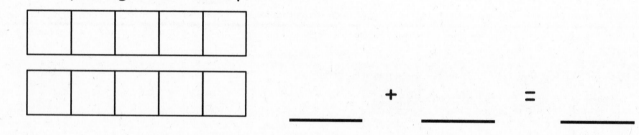

_____ + _____ = _____

Color 0 squares green and 1 square blue.

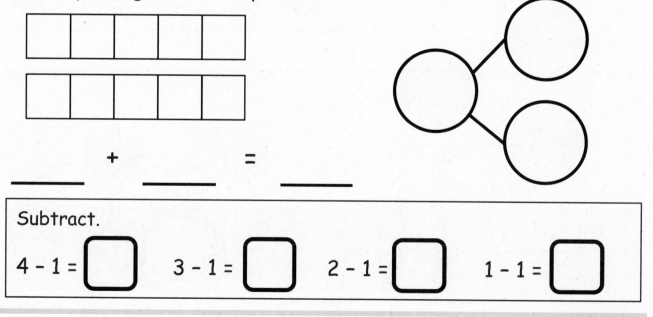

_____ + _____ = _____

Subtract.

$4 - 1 =$ ☐ $3 - 1 =$ ☐ $2 - 1 =$ ☐ $1 - 1 =$ ☐

Lesson 38: Add 1 to numbers 1–9 to see the pattern of *the next number* using 5-group drawings and equations.

Name _____   Date _____

Draw dots to make 10. Fill in the number bond.

Lesson 39:   Find the number that makes 10 for numbers 1–9, and record each
with a 5-group drawing.

©2015 Great Minds. eureka-math.org
GK-M4-SE-B3-1.3.1-01.2016

163

Draw dots to make 10. Fill in the number bond.

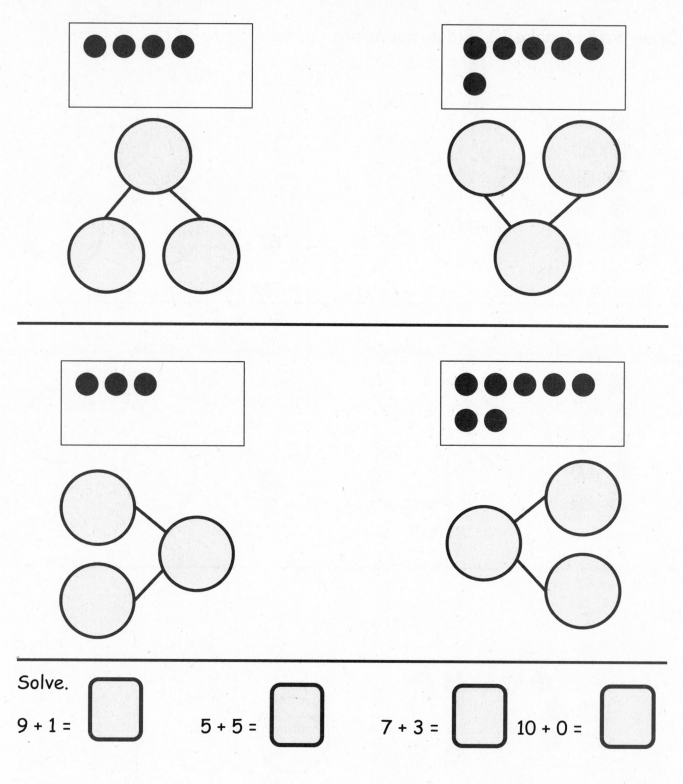

Solve.

9 + 1 =   5 + 5 =   7 + 3 =   10 + 0 =

Lesson 39:   Find the number that makes 10 for numbers 1–9, and record each
with a 5-group drawing.

Name _____   Date _____

Draw dots to make 10. Finish the number bonds. Draw a line from the 5-group to the matching number bond.

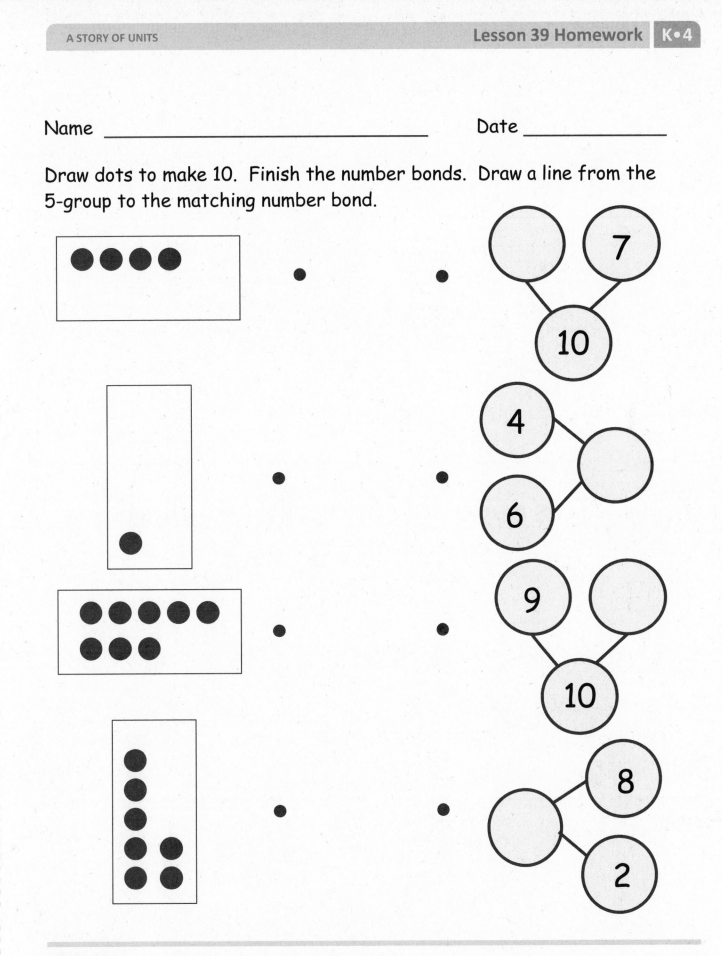

Lesson 39:   Find the number that makes 10 for numbers 1–9, and record each
with a 5-group drawing.

©2015 Great Minds. eureka-math.org
GK-M4-SE-B3-1.3.1-01.2016

This page intentionally left blank